Service Life Analysis of Water Main Epoxy Lining

Awwa Research Foundation

About the Awwa Research Foundation

The Awwa Research Foundation (AwwaRF) is a member-supported, international, nonprofit organization that sponsors research to enable water utilities, public health agencies, and other professionals to provide safe and affordable drinking water to consumers.

The Foundation's mission is to advance the science of water to improve the quality of life. To achieve this mission, the Foundation sponsors studies on all aspects of drinking water, including supply and resources, treatment, monitoring and analysis, distribution, management, and health effects. Funding for research is provided primarily by subscription payments from approximately 1,000 utilities, consulting firms, and manufacturers in North America and abroad. Additional funding comes from collaborative partnerships with other national and international organizations, allowing for resources to be leveraged, expertise to be shared, and broad-based knowledge to be developed and disseminated. Government funding serves as a third source of research dollars.

From its headquarters in Denver, Colorado, the Foundation's staff directs and supports the efforts of more than 800 volunteers who serve on the board of trustees and various committees. These volunteers represent many facets of the water industry, and contribute their expertise to select and monitor research studies that benefit the entire drinking water community.

The results of research are disseminated through a number of channels, including reports, the Web site, conferences, and periodicals.

For subscribers, the Foundation serves as a cooperative program in which water suppliers unite to pool their resources. By applying Foundation research findings, these water suppliers can save substantial costs and stay on the leading edge of drinking water science and technology. Since its inception, AwwaRF has supplied the water community with more than $300 million in applied research.

More information about the Foundation and how to become a subscriber is available on the Web at **www.awwarf.org.**

Service Life Analysis of Water Main Epoxy Lining

Prepared by:
Arun K. Deb, **Jerry K. Snyder**, **John O. Hammell, Jr.**, and **Elizabeth Tyler**
Weston Solutions, Inc., West Chester, Pa. 19380

Linda Gray
KTA-Tator (Canada), Inc.

and

Ian Warren
WRc-NSF Ltd., United Kingdom

Sponsored by:
Awwa Research Foundation
6666 West Quincy Avenue, Denver, CO 80235-3098

Published by:

Awwa Research Foundation

American Water Works Association

IWA Publishing

DISCLAIMER

This study was funded by the Awwa Research Foundation (AwwaRF). AwwaRF assumes no responsibility for the content of the research study reported in this publication or for the opinions or statements of fact expressed in the report. The mention of trade names for commercial products does not represent or imply the approval or endorsement of AwwaRF. This report is presented solely for informational purposes.

CONTENTS

TABLES

FIGURES

FOREWORD

The Awwa Research Foundation is a nonprofit corporation that is dedicated to the implementation of a research effort to help utilities respond to regulatory requirements and traditional high-priority concerns of the industry. The research agenda is developed through a process of consultation with subscribers and drinking water professionals. Under the umbrella of a Strategic Research Plan, the Research Advisory Council prioritizes the suggested projects based upon current and future needs, applicability, and past work; the recommendations are forwarded to the Board of Trustees for final selection. The foundation also sponsors research projects through the unsolicited proposal process; the Collaborative Research, Research Applications, and Tailored Collaboration programs; and various joint research efforts with organizations such as the U.S. Environmental Protection Agency, the U.S. Bureau of Reclamation, and the Association of California Water Agencies.

This publication is a result of one of those sponsored studies, and it is hoped that its findings will be applied in communities throughout the world. The following report serves not only as a means of communicating the results of the water industry's centralized research program, but also as a tool to enlist the further support of the nonmember utilities and individuals.

Projects are managed closely from their inception to the final report by the foundation's staff and large cadre of volunteers who willingly contribute their time and expertise. The foundation serves a planning and management function and awards contracts to other institutions such as water utilities, universities, and engineering firms. The funding for this research effort comes primarily from the Subscription Program, through which water utilities subscribe to the research program and make an annual payment proportionate to the volume of water they deliver, and consultants and manufacturers subscribe based on their annual billings. The program offers a cost-effective and fair method for funding research in the public interest.

A broad spectrum of water supply issues is addressed by the foundation's research agenda: resources, treatment and operations, distribution and storage, water quality and analysis, toxicology, economics, and management. The ultimate purpose of the coordinated effort is to assist water suppliers to provide the highest possible quality of water economically and reliably. The true benefits are realized when the results are implemented at the utility level. The foundation's trustees are pleased to offer this publication as a contribution toward that end.

Walter J. Bishop
Chair, Board of Trustees
Awwa Research Foundation

Robert C. Renner, P.E.
Executive Director
Awwa Research Foundation

ACKNOWLEDGMENTS

The authors of this report thank the Awwa Research Foundation (AwwaRF) for its support of this study, and the AwwaRF project manager, Dr. Jian Zhang, for his help and guidance in the execution of the project. The authors acknowledge the input provided by the members of the Project Advisory Committee (PAC), namely:

- Heitkamp, Incorporated, Watertown, Conn., Mr. Benedict H. Ebner
- City of Cleveland, Division of Water, Cleveland, Ohio, Mr. Stanley Dobrowski
- Bucks County Water and Sewer Authority, Warrington, Pa., Mr. John Butler

The participating utilities include:

- City of Ottawa Drinking Water Services, Ottawa, Cana., Mr. Robert Cowan
- Chester Water Authority, Chester, Pa., Mr. Brian McEwen
- Bucks County Water and Sewer Authority, Warrington, Pa., Mr. John Butler
- Denver Water, Denver, Colo., Mr. John H. Bambei, Jr.
- Halifax Regional Water Commission, Halifax, Cana., Mr. Jamie Hannam
- Severn Trent Water Ltd., Birmingham, United Kingdom, Mr. Michael J. Kane
- Yorkshire Water Ltd., Bradford, United Kingdom, Mr. David Brown
- Thames Water Plc., Reading, United Kingdom, Dr. David M. Holt

SINTEF Applied Chemistry of Norway, Mr. Leif Sigurd Hafskjold; Raven Lining Systems, Tulsa, Okla., Mr. Stewart Nance; and Mainlining Service, Inc., Elma, N.Y., Mr. Craig Perkins provided information on epoxy lining construction.

The authors would also like to thank the workshop participants for their input.

EXECUTIVE SUMMARY

Research was conducted to assess the long-term durability of epoxy resin lining for use in lining potable water mains. For old potable water mains in good structural condition, but in need of rehabilitation, cleaning and lining is the most economical option for rehabilitation of these pipes. Many different cleaning and lining techniques and materials are available; the effectiveness and useful life of which vary depending on the physical properties of the liner, the construction methods, and the corrosivity of the water. In North America, most water main rehabilitation efforts have been conducted using cement mortar lining, which has been found to be ineffective as a long-term solution when water is relatively soft and aggressive. An alternative to cement mortar lining is epoxy resin lining, which has been found to have high durability and higher hydraulic capacity, particularly in soft aggressive waters.

Due to North America's limited experience with this practice, the Awwa Research Foundation (AwwaRF) contracted Weston Solutions, Inc. to review the historical performance of epoxy lining of water mains and to conduct accelerated tests to assess the longevity of epoxy liners. This study gives water utilities adequate information on epoxy linings to control the lining construction process and to assess its long-term performance. The useful life of epoxy liners was estimated to be approximately 40 to 60 years, depending on the use of proper construction methods. Water utilities in North America should now consider epoxy lining as a cost-effective alternative for water main rehabilitation.

BACKGROUND INFORMATION

As water mains become old and reach the end of their useful lives, their performance diminishes gradually, resulting in high maintenance costs, deterioration of water quality, loss of hydraulic capacity, and a significant increase in customer complaints. About 50% of the water mains in North American water systems are of cast iron. Most of the cast iron mains installed prior to the 1950s are unlined. Many of these water mains are structurally sound. Cleaning and lining of such a structurally sound, but otherwise poorly performing, water main will decrease customer complaints and will significantly improve water quality, flow, and pressure. This will improve the long-term performance of the pipe and improve the reliability of the system. Rehabilitation of water mains using cleaning and lining results in significant cost savings compared to replacement.

In North America, most of the water main rehabilitation has been conducted using cement mortar lining. An alternative to cement mortar lining of water mains is epoxy resin lining. The experience with epoxy lining of water mains in the United Kingdom, and an AwwaRF study in North America in 1993 (Conroy et al., 1993), confirmed the viability of epoxy lining for water main rehabilitation.

Epoxy resin lining provides a corrosion-resistant lining system in pipes that improves flow capacity and does not affect water quality. All epoxy resin material used for in situ lining of potable water mains must be (AWWA, 2002):

- Solvent- and benzyl-alcohol free
- High build
- Slump-resistant
- Rapid-curing
- Moisture-tolerant

The basic principles of epoxy resin formulation are well documented. Resin material types are generally, but not universally, classified as being Type 1 or Type 2 materials. Type 1 materials are typified by the early formulations and are generally low viscosity materials that can be pumped by ordinary lining equipment. Type 2 resins are more recent. These are higher viscosity materials requiring heated lining hoses.

In situ epoxy resin lining of water mains was introduced in the United Kingdom in 1979 (Warren, 1983). As this new technology was developed, it became more widely used, particularly in aggressive waters where previously applied lining methods, such as cement mortar lining, had failed (Warren, 1983). Epoxy resin lining techniques offer a cost-effective way to rehabilitate potable water pipes and have been used successfully in the United Kingdom, Japan, and, to a limited extent, in the United States (Salvo, 1998; Warren, 1996).

A standard quality assurance and quality control (QA/QC) procedure for installation of liners is very important for the longevity of liners. Ongoing monitoring should be conducted to determine the effectiveness of the lining. Epoxy lining became widely used in the United Kingdom, but there were no national standards or guidelines published until WRc developed a work method in 1989 (WRc, 1989). Prior to 1989, individual water utilities controlled the application techniques, which led to a lack of consistency between lining applications. The American Water Works Association (AWWA) is developing a draft standard for *Spray Applied In Situ Epoxy Lining of Water Pipelines* (AWWA, 2002). In the United Kingdom, WRc published, most recently, the *In Situ Epoxy Resin Lining—Operational Requirements and Code of Practice* (Code of Practice) (WRc, 2001).

In addition to following QA/QC procedures for the installation of epoxy lining, it is also required that the epoxy resin material be certified. In the United States, the material must be certified according to American National Standards Institute/National Sanitation Foundation (ANSI/NSF) Standard 61 (NSF 61).

OBJECTIVES OF THE RESEARCH

The specific objectives of this research were to:

- Gather information on historical epoxy lining installations, and document the effectiveness of these previously installed liners in protecting cast iron or steel water mains.
- Develop protocols and procedures, and conduct accelerated life-cycle tests on epoxy liners for their assessment of long-term performance.
- Assess the longevity of early epoxy liner installations for renewal planning.

RESEARCH APPROACH

The technical approach to accomplish the project objectives and goals consisted of the following tasks:

- A comprehensive review of available information was conducted to identify the status of epoxy lining materials, technologies, and quality controls, and to identify water utilities with historically installed epoxy liners.
- Questionnaire surveys were conducted to determine the extent to which epoxy resin is used in lining potable water mains, to identify current lining practices, and to survey

epoxy manufacturers about currently manufactured epoxies and their properties. Selected groups of water utilities with experience with epoxy liners and epoxy manufacturers were contacted and questionnaires were distributed.

- Based on the results of the questionnaires, an in situ lining sampling plan was developed to exhume pipe coupons of various lining ages and types of epoxies. Coupons of historical water main pipes with epoxy linings were collected from various utilities in the United Kingdom, the United States, and Canada. All samples were transported to the KTA-Tator Laboratory in Canada. These coupons were physically inspected, photographed, and documented. The historically epoxy-lined pipes were analyzed for the condition of epoxy linings with respect to the ages of linings, and the test results were summarized.
- Results were analyzed and principal parameters affecting the performance of epoxy liners were identified.
- Long-term accelerated test methodologies were developed with the goal of developing a method to predict the performance of epoxy linings. The testing methodologies were based on a review of existing long-term accelerated testing methods and protocols for preparation of test samples.
- Long-term accelerated tests were conducted on four different epoxy samples and one polyurethane sample. The results were analyzed and important parameters were identified.
- A methodology was developed for estimating the useful life of epoxy linings and subsequently, the useful life of epoxy lining was estimated.
- An industry workshop was conducted to review the project tasks and validate the procedure developed for estimating the useful life of epoxy lining.
- Findings and conclusions were drawn, and recommendations were made for use by water utilities. Based on these tasks, the final report was prepared.

SURVEY OF CURRENT PRACTICES

A questionnaire was distributed to various water utilities in North America and the United Kingdom to survey the extent to which these utilities have performed cleaning and epoxy lining in their systems. The water utility questionnaire requested information about the reasons why epoxy liners were used, the performance of in situ liners, and the installation and QA/QC procedures used by the utility. Other basic information included:

- Year of installation of lining
- Pipe diameter
- Length of pipe lined
- Epoxy manufacturer
- Epoxy product
- Contractor
- Cleaning method used
- Year pipe was installed
- Pipe material

Results of the questionnaire indicated that North American water utilities installed epoxy lining in about 100,000 feet of water mains, whereas the United Kingdom utilities installed epoxy lining in about 50,000,000 feet of water mains. Most of the water mains lined with epoxy are small diameter (3 to 12 inches) unlined cast iron water mains. Water quality consideration is found to be the primary reason for lining water mains. Significant improvement in water quality and hydraulic performance was observed by water utilities after installation of liners. All utilities have a good written QA/QC procedure and conduct closed-circuit television (CCTV) inspection of water mains before and after lining.

The responses received from the questionnaire were used to develop a matrix of the types of epoxy materials used and the years in which they were applied. A pipe sampling program was subsequently developed from this matrix with the objective of collecting samples of epoxy-lined pipes covering a wide range of epoxy materials and ages.

Another questionnaire was prepared and sent to epoxy manufacturers to obtain the properties of seven currently available epoxy materials and one polyurethane formulation. Significant differences in material properties were shown.

ANALYSIS OF HISTORICAL EPOXY-LINED PIPES

Historical epoxy-lined pipe samples were collected from five water utilities in North America and the United Kingdom in order to analyze conditions of historical epoxy linings, to determine important parameters affecting the longevity of the liners, and to estimate the longevity of such liners. The 5 utilities provided 17 epoxy-lined pipe samples with 8 different epoxy materials of various ages, from 1 to 25 years. These samples were sent to KTA-Tator Laboratory in Edmonton, Cana.

The samples were analyzed for dry film thickness (DFT), holidays (small pinholes), electrochemical impedance spectroscopy (EIS) measurements, knife adhesion, blister evaluation, observations or rust rating, visual observations regarding application of the coating and condition of the underlying steel surface, and observations by stereomicroscopy, as applicable.

In general, epoxy linings were found to be in good to excellent condition, with the exception of three samples. Two of these samples were old and were not constructed with proper QA/QC.

Data collected on various parameters from laboratory analysis of field pipe lining samples were analyzed to identify important parameters affecting the performance of epoxy liners. The following parameters appeared to be the most important:

- EIS
- DFT
- Holiday density
- Knife adhesion

The barrier property of a liner, which is dependent on the properties of the epoxy material, is measured by EIS and is perhaps the most important parameter for the effectiveness and longevity of the liner. Based on EIS measurements, all of the epoxy linings maintained excellent barrier properties, except for the two highly deteriorated linings. The excellent barrier properties indicate that the liners have experienced minimal physical or chemical deterioration. Further, it indicates that the rate of water diffusion through the liner is very low and that the epoxy liner provides good protection to the metal substrate.

The DFT of the epoxy linings varied widely between samples and within a sample. Variations within a sample were typically the result of irregularities in the epoxy application. Continuous application of epoxy lining is needed for a constant DFT along the length of a potable water main.

Holidays in a liner, which are small pinholes, are found to be one of the most important parameters for performance and longevity of liners and are generally formed due to a lack of proper construction controls. Many of the epoxy linings had holidays, which ranged in size from microscopic to about 1 millimeter (mm) in size. The holidays presumably formed from surface contaminants, surface irregularities, or spray-related problems.

Even if the epoxy material is good, poor installation of the liner may result in holidays, thereby allowing water to come in contact with the metal pipe and reduce the overall effectiveness of the liner. Therefore, construction control of epoxy linings is very important to achieve a durable liner with a long life expectancy. For a well-constructed lining, holiday density should be close to zero.

Knife adhesion is a parameter that indicates the adhesion of the lining to the wall of the water pipe. Knife adhesion depends largely on the preparation of the surface and also on the epoxy material. Based on knife adhesion, all of the linings had low adhesion to the metal surface.

Adherence to good construction practices, which affect holiday density, DFT, and knife adhesion, and high barrier properties measured by EIS lead to high-quality epoxy liners that have the potential to extend the life expectancy of a potable water main.

ACCELERATED TESTS

In order to estimate the longevity of epoxy linings, accelerated tests were conducted using four different types of epoxy linings and one rapid-set polyurethane lining. The five linings were applied at target film thicknesses of 40 mils. Two of the epoxies were also applied at a target film thickness of 20 mils, to represent a compromised situation under which more rapid deterioration should occur.

Steel test panels 3 x 6 inches in size were lined. A total of 80 panels were prepared for each combination of lining and film thickness. In order to find the impact of surface preparation, the surfaces of the panels were prepared in two different ways. Side A ("front" side) of each panel was solvent-cleaned, followed by blast-cleaning to a near white metal blast, whereas side B was solvent-cleaned only. The test panels were immersed in deionized water at temperatures 8 degrees Celsius (°C), 25 °C, and 35 °C. The panels were evaluated at 0, 1-, 2-, 4-, and 8-month intervals for knife adhesion, pull-off adhesion, EIS, film thickness, and Buchholz hardness.

Knife adhesion was poor on non-blasted surfaces. Adhesion was initially high on blast-cleaned surfaces, but dropped with time. Impedance (measured using EIS) was generally high, indicating good barrier properties even at 35 °C. A longer test duration is required for detectable deterioration by EIS.

LONGEVITY OF EPOXY LINING

Results of this research show the estimated useful life of epoxy resin linings to be approximately 40 to 60 years. To estimate the useful life of the linings, a methodology was developed using the EIS measurements taken from the immersion panels and the historical samples of epoxy-lined pipes obtained from various water utilities.

The EIS measurements for the immersion panels were plotted on the same graph as the historical pipe samples for the same epoxy. This graph representing EIS values with time of installation showed a decreasing trend for which a regression line was drawn. The regression line was extrapolated to EIS values of 6 and 4, which represent the end of a liner's useful life and total liner failure, respectively. Using this regression analysis, the useful life of an epoxy liner was estimated to be approximately 123 years and total failure was estimated to occur after 172 years. This service life is based on the ideal condition of construction with strict adherence to QA/QC procedures. Considering a factor of safety of 3 based on construction irregularity, the service life of epoxy lining was estimated to be approximately 40 to 60 years.

FINDINGS AND RECOMMENDATIONS

The following are the important findings and recommendations of this study:

- Water quality degradation is a key reason for lining water mains. Improvements in water quality are noted subsequent to lining installation.
- Only a few North American water utilities have experience in epoxy lining of potable mains.
- For properly constructed epoxy linings, water utilities may expect a lining life of 40 to 60 years.
- Controlled surface preparation and adherence to liner application methodologies are the most critical factors in ensuring an effective and long-lasting liner.
- Impedance, DFT, adhesion, and holiday density are important parameters that affect the longevity of a liner.
    - Impedance measurements are a good indication of the barrier properties of liner materials; high impedance values indicate a good barrier.
    - Impedance can be measured accurately using EIS technology, and water utilities should consider measuring impedance to understand the barrier property of a liner.
    - Historical epoxy liners with high impedance values and low holiday densities were found to be in excellent condition.
    - Corrosion of the steel surface under the coating was related to barrier properties and holiday density.
    - DFT is an important parameter for holidays and under-film corrosion.
- Epoxy lining of water mains has been found to be a good alternative to cement mortar lining and all North American water utilities should consider this technology as an alternative to cement mortar lining.
- Water utilities should adapt and use proper QA/QC procedures in order to obtain good longevity of the lined water mains.

CHAPTER 1
INTRODUCTION

BACKGROUND AND INTRODUCTION

As water mains become old and reach the end of their useful lives, their performance diminishes gradually, resulting in high maintenance costs, deterioration of water quality, loss of hydraulic capacity, and a significant increase in customer complaints. With the rehabilitation of a poorly performing water main using cleaning and lining, maintenance costs and customer complaints will decrease, and flow, pressure, and water quality will improve significantly. This will improve the long-term performance of the pipe and improve the reliability of the system.

Over 80% of water utility investment is committed to the distribution/transmission system. Water systems in many communities and cities in North America are old. In order to rehabilitate and renew the water distribution systems, the American Water Works Association (AWWA) recently estimated that approximately $325 billion (AWWA, 1999) is needed over the next 20 years. The distribution system is the last link between a water utility and its customers; therefore, management and proper maintenance of the distribution mains play a key role in water utility service to customers.

About 50% of water mains in North American water systems are of cast iron. Most of the cast iron mains installed prior to the 1950s are unlined. Many of these water mains are structurally sound, but show severe tuberculation, resulting in reduction of hydraulic capacity and water quality problems. The Hazen-Williams "C" values of these pipes have been reduced significantly, resulting in high energy costs. Many of these pipes may be structurally in good condition. If the water main is structurally in good condition, cleaning and lining is the most economical option for rehabilitation of these pipes; however, if the main is not in structurally good condition, either due to graphitization or external corrosion, replacement of the pipe will be the desired option.

Many different pipe cleaning and lining techniques are available. The effectiveness and life of these technologies vary depending on the pipe material, purpose, corrosivity of water, and lining material. Rehabilitation of water mains using cleaning and lining results in significant cost savings compared to replacement; however, if the proper technology is not used, the life of rehabilitated pipe may be reduced significantly. In North America, most of the water main rehabilitation was conducted using cement mortar lining.

Alternatives to cleaning and cement mortar lining of water mains were studied, and in 1993 the Awwa Research Foundation (AwwaRF) conducted a demonstration study using the epoxy-lining technique in the United States (Conroy et al., 1993). This demonstration confirmed the viability of epoxy lining for water main rehabilitation.

Epoxy lining of cast iron and steel pipelines was first introduced in the United Kingdom (Warren, 1989) in the late 1970s, and is considered a proven process for protection of an interior pipeline wall from corrosion. In the United Kingdom, cement lining was found to be ineffective as a long-term solution when water is relatively soft and aggressive. As an alternative to cement mortar lining, epoxy lining has been found to have high durability and higher hydraulic capacity. In 1985, approval of epoxy by the United Kingdom Department of Environment for use in public water supply resulted in an accelerated growth in the use of epoxy lining in water mains. Since then, thousands of kilometers of pipes have been lined using the epoxy lining technique in the United Kingdom. Now epoxy lining for rehabilitation of water mains from 3 inches to 12 inches in diameter is most common in the United Kingdom and Japan. In soft water areas, this technology is used exclusively. The thickness of the epoxy lining is about 1 millimeter (mm), whereas the thickness of cement mortar lining is about 6 mm, which results in epoxy lining providing a larger area and flow.

In the United States, however, the first test case demonstration of epoxy lining of water mains was conducted by AwwaRF in 1993 (Conroy et al., 1993), by lining a 6-inch unlined cast iron pipe at the Chester Water Authority, in Pennsylvania. Water quality problems as a result of leaching chemicals from the epoxy lining was a big concern in the United States. The AwwaRF study showed that leaching from epoxy lining is minimal and satisfies all primary drinking water regulations. However, without the approval of the National Sanitation Foundation (NSF), water utilities were very hesitant to accept epoxy lining as an alternative to cement lining. A few years later, NSF provided the approval certification (NSF 61) for use of epoxy as a viable alternative to cement mortar lining for drinking water mains. Recently, AWWA also developed a draft standard for in situ epoxy lining of water pipelines. As a result, there is an impetus for North American water utilities to consider the epoxy-lining technique as an alternative to cement lining.

Because the useful life of each lining material may vary, and also may depend on the quality of water and protection provided, when comparing the cost of alternative cleaning and lining technologies available for water main rehabilitation, it is necessary to consider the useful life of the lining material. Due to a delay in the NSF approval process for application of epoxy lining in the United States, water utilities in the United States do not have long-term experience with epoxy lining of water mains. This project analyzed protocols and procedures that will help water utilities to assess the long-term performance of epoxy liners.

North American water utilities will benefit greatly by the use of this information to identify the suitability of epoxy liners in cost-effective implementation of their water main rehabilitation programs. Water utilities in North America are now faced with extensive rehabilitation and replacement of water mains, spending billions of dollars every year. In order to manage water distribution infrastructure costs effectively, it is necessary to consider epoxy lining as an alternative to cement-mortar lining for analysis in the decision process. This study provides appropriate guidance to water utilities in determining the longevity of epoxy lining under various conditions. This study helps water utilities to conduct analysis of epoxy lining and assess long-term performance in order to determine the most cost-effective and suitable lining technology that will solve the problems. In addition, this research provides water utilities with a methodology to identify important parameters and to estimate the service life of epoxy liners to assess their long-term performance.

OBJECTIVES OF THE RESEARCH

The objectives of this research were the following:

- Gather information on historical epoxy-lining installations, and document the effectiveness of these previously installed liners in protecting cast iron or steel water mains.
- Develop protocols and procedures to conduct accelerated life-cycle tests on epoxy liners for their assessment of long-term performance.
- Assess the longevity of epoxy liner installations for renewal planning.

RESEARCH APPROACH

The technical approach to accomplish the project objectives and goals consisted of the following tasks:

- A comprehensive review of available information was conducted to identify the status of epoxy lining materials, technologies, and quality controls, and to identify water utilities with historically installed epoxy liners.
- Questionnaire surveys were conducted to determine the extent to which epoxy resin is used in lining potable water mains, to identify current lining practices, and to contact epoxy manufacturers regarding currently manufactured epoxies and their properties. Selected groups of water utilities with experience using epoxy liners and epoxy manufacturers were contacted and questionnaires were distributed.
- Based on the results of the questionnaires, an in situ lining sampling plan was developed to exhume pipe coupons of various lining ages and types of epoxies. Coupons of water main pipes with epoxy linings were collected from various utilities in the United Kingdom, the United States, and Canada. All samples were transported to the KTA-Tator Laboratory in Canada. These coupons were physically inspected, photographed, and documented. The historically epoxy-lined pipes were analyzed for the condition of epoxy linings with respect to the ages of linings, and the test results were summarized.
- Results were analyzed and principal parameters affecting the performance of epoxy liners were identified.
- Long-term accelerated test methodologies were developed with the goal of developing a method to predict the performance of epoxy linings. The testing methodologies were based on a review of existing long-term accelerated testing methods and protocols for preparation of test samples.
- Long-term accelerated tests were conducted on four different epoxy samples and one polyurethane sample. The results were analyzed and important parameters were identified.
- A methodology was developed for estimating of the useful life of epoxy linings and subsequently, the useful life of epoxy linings was estimated.
- An industry workshop was conducted to review the project tasks and validate the procedure developed for estimating of the useful life of epoxy linings.
- Findings and conclusions were drawn, and recommendations were made for use by water utilities. Based on these tasks, the final report was prepared.

ORGANIZATION OF REPORT

Chapter 1 provides the background information, objectives, and scope of this report. Chapter 2 summarizes the literature review, and analysis to identify current epoxy lining operational practices. Chapter 3 summarizes questionnaire data collected from water utilities and from epoxy manufacturers to identify water utilities with experience in epoxy lining and various epoxies available on the market for lining of water mains. Chapter 4 describes the collection, testing, and analysis of pipe lining samples from historical epoxy-lined pipes. Chapter 5 provides a description of parameters that affect the performance of epoxy liners. Chapter 6 describes test methodologies, preparation of samples, test procedures, quality control (QC), and analysis of results and identification of important parameters. In Chapter 7, historical performance data, along with accelerated test data, were used to develop a methodology for estimation of the useful life of epoxy linings. Chapter 8 presents the findings and recommendations of this project.

CHAPTER 2
REVIEW OF THE CURRENT INFORMATION AND LITERATURE

LITERATURE REVIEW

A literature review was conducted to establish the materials, practices, and applications used in the epoxy resin lining processes. This literature review was part of an effort to develop a protocol and procedure that will help water utilities assess the long-term performance of epoxy resin liners applied to water mains. Epoxy resin lining is a technique used in nonstructural rehabilitation of potable water mains.

Previous research efforts reviewed for this literature survey included articles, guidelines published for epoxy lining operations, and case studies. The sources reviewed conveyed information relating to the ways in which epoxy resin lining has been applied in the past and how linings have performed. The evolution of epoxy lining was assembled, the characteristics of materials were identified, and the performance of this type of lining was evaluated. The life cycle of a rehabilitated pipe was assessed by considering the useful life of epoxy resin lining materials used in pipe rehabilitation efforts.

The applicability and durability of epoxy resin lining as used for nonstructural rehabilitation of potable water mains was assessed. Quality assurance and quality control (QA/QC) issues were investigated, as well as short- and long-term monitoring efforts. Monitoring efforts play a role in QA/QC and in determining the effects of an epoxy-lined pipe on water quality and hydraulic performance. Information was also gathered from case studies on historical epoxy lining installations to analyze the effectiveness of previously installed liners in protecting potable water mains.

PIPE LINING TECHNIQUES

Many potable water systems are comprised of pipes that are old and in need of repair. In general, these pipes are made of cast iron, are unlined, and are structurally sound, but exhibit signs of internal corrosion, such as tuberculation or deterioration in water quality and loss of hydraulic capacity (Warren et al., 1996). Pipe-lining techniques were first developed to remedy such situations by cleaning the corrosion buildup and providing the interior of pipes with a corrosion-resistant surface lining and thus, allowing the life of the pipes to be extended.

The effectiveness and life of different pipe-lining technologies vary depending on the pipe material, purpose, corrosivity of water, and lining material. One method of pipe rehabilitation involves cleaning sections of pipe and lining them with cement mortar. In North America, Portland cement mortar lining is the conventional choice to renovate unlined cast iron pipe for mains 4 inches (100 mm) and larger in diameter. In many cases, this type of lining has proved to be effective, but studies and field experience have shown that, in soft waters, the lining may corrode and adversely affect water quality for a sustained period as it deteriorates. In the United Kingdom, cement lining was used from the 1930s through the 1990s but was found to be ineffective as a long-term solution when water is relatively soft and aggressive, thus corroding the cement lining. Bitumen lining was introduced in 1965 as an alternative to cement mortar lining, but this material proved to be ineffective as a durable liner for potable water mains. Use of bitumen linings was discontinued in 1980. Epoxy resin lining was introduced as a cost-effective alternative and it eventually replaced cement mortar and bituminous lining techniques in the United Kingdom (WRc, 1997).

In the United Kingdom, epoxy resin lining techniques were first introduced in 1979 and were widely used and accepted for lining potable water pipelines. Research has been conducted to improve materials for lining applications and to develop operational practices designed to aid in using epoxy-lining techniques as a method of nonstructural pipe rehabilitation. These operational guidelines set the standard that contractors in the United Kingdom must achieve in order to participate in epoxy-lining procedures (WRc, 1997). However, currently, epoxy resin lining is not widely used throughout the United Kingdom. The majority of water utilities have chosen to use polymeric coatings instead of epoxy resin. Although polymeric coatings have only been used in recent history, operational standards and guidelines have been published and the use of polymeric coatings has been widely accepted throughout the United Kingdom (Warren Associates, 2000).

To line potable water pipes with any material, the material must be certified according to American National Standards Institute/National Sanitation Foundation (ANSI/NSF) Standard 61 (NSF 61) (AWWA, 2002) in the United States, and Drinking Water Inspectorate Regulation 31 in the United Kingdom. Both regulations ensure acceptable quality for water that comes in contact with any substance. ANSI/NSF 61 covers indirect additives, products, and materials that must be tested for effects on water quality and their fitness for purpose (WRc-NSF, 2004). By 1998, five epoxy resin products had been approved under this standard in North America (Salvo, 1998).

DEFINITION OF EPOXY RESIN LINING

Epoxy resin is a type of material used in nonstructural pipe rehabilitation. Epoxies consist of two components that react to form a hard inert material. The first component consists of an epoxy resin and the second consists of an epoxy curing agent, also called a hardener. Employing different types and amounts of hardener tends to control the cross-link density between the resin and hardener, thus, varying the structure of the material (Bhangale, 2003). Different structures yield different material characteristics, such as strength or hardness. Therefore, it is important to choose epoxy resin materials carefully and to ensure that the mix ratio between the resin and hardener components is correct and carefully monitored during lining processes (WRc, 1997). The correct mix ratio of resin to hardener varies depending on the epoxy used. Epoxy resin manufacturers specify the optimum mix ratio required to achieve a lining that exhibits good adhesion to the interior surface of the pipe, resistance to moisture, minimum leaching potential, and adequate hardness (WRc, 1997).

All epoxy resin material used for in situ lining of potable water mains must be (AWWA, 2002):

- Solvent and benzyl alcohol free
- High build
- Slump-resistant
- Rapid-curing
- Moisture-tolerant

The basic formulations are all very similar because of the limitation put on the formulators when developing materials that will cure within 16 hours at temperatures as low as 3 degrees Celsius (°C), while providing water quality acceptability (DWI, 2005). General information on the formulations is in the public domain, while the specific individual formulations are proprietary. The basic principles of epoxy resin formulation are well-documented. Resin material types are, generally, but not universally, classified as being Type 1 or Type 2 materials.

Type 1 materials are typified by the early formulations and are generally low viscosity materials that can be pumped by ordinary hose lining equipment and have a curing agent based on trimethyl hexamethylene diamine (AWWA, 2003). Table 2.1 shows typical Type 1 formulations (AWWA, 2002).

Type 2 resins are more recent and are higher viscosity materials requiring heated lining hoses and rigs for application and have curing agents based on more complicated alternative amine systems. Type 2 formulations are very similar to the above for the resin component, but in the hardener, the amine curing agents are more complex. One very successful curing agent was based on amines derived from cashew nut shell liquid, but there are a variety of alternatives available (AWWA, 2002).

Table 2.1
Typical Type 1 formulation

Resin (base)	Hardener (catalyst)
Bisphenol A/F diglycidyl ether	Trimethyl hexamethylene diamine
Reactive diluents	Reactive diluents
Inert China clay fillers	Inert China clay fillers
Thixotropes	Thixotropes
Pigments (iron oxide)	Pigments

To be suitable for pipe lining, epoxy resin must satisfy the following requirements (Warren, 1983):

- Be suitably applicable to wet surfaces
- Cure rapidly to allow return to service after 16 hours of air curing, thus minimizing consumer disruption
- Produce a smooth, pinhole-free lining
- Have long-term durability
- Not impair water quality

Epoxy resin lining provides a corrosion-resistant lining system in pipes that improves flow capacity and does not affect water quality. For a structurally sound water main, application of epoxy resin lining is a more efficient and less expensive rehabilitation method than replacement. To successfully use epoxy lining, the pipe section needing repair must be cleaned according to recommended methods and the carefully chosen epoxy material must be applied as per published guidelines (WRc, 1997). In addition, testing and monitoring should be implemented before, during, and after application processes as part of QA/QC. Epoxy resin lining techniques have been used successfully in the United Kingdom, Japan, and, to a limited extent, in the United States.

The epoxy resin materials that have been currently available and are in use in the United Kingdom for in situ water main lining are summarized in Table 2.2 (DWI, 2005).

Table 2.2
Recognized epoxy resins suitable for use in coating pipes in the United Kingdom

Company	Material
E. Wood Ltd	Copon Hycote 162 PWX
Fosroc Ltd	Nitoline WP
	Nitoline WP 2
Hunting Industrial Coatings Ltd	Hunting Waterline
Mercol Descaling Co. Ltd	Geopox GX014
Subterra	ELC 173/90
W&J Leigh & Co.	Pipegard P300

EVOLUTION OF EPOXY RESIN LINING

In situ epoxy resin lining of water mains was introduced in the United Kingdom in 1979 (Warren, 1983). As this new technology was developed, it became more widely used, particularly in aggressive waters where previously applied lining methods, such as cement mortar lining, had failed (Warren, 1983). Epoxy resin lining techniques offer a cost-effective way to rehabilitate potable water pipes and extend their service lives by an estimated 30 to 50 years by providing a protective barrier between the water and the host pipe (Salvo, 1998; Warren, 1996).

An article, published in *Chemistry and Industry* entitled "Factors Influencing the Choice of Materials for Epoxy Resin Renovation of Water Mains," outlines the evolution of epoxy resin lining used for potable water mains (Warren, 1983). Epoxy resin lining was tested for use in potable water mains because of a need to find an alternative to cement mortar lining, particularly in the United Kingdom. Testing began with epoxy resins developed for underwater repair work on North Sea oil rigs. The life of epoxy resins for use in potable water pipes was researched by conducting accelerated testing, which involved placing samples in rigs that contained aerated water dosed with chloride and sulfate at 150 milligrams per liter (mg/L) at 50 °C (Warren, 1983). Therefore, the corrosion rate was accelerated and the durability of different epoxy resin materials was tested.

Following adequate results of laboratory-based accelerated testing, in-service assessments ensued, which involved lining pipe samples with different epoxy resins and installing inspection pits in various distribution systems covering a range of pipe diameters and water qualities. Over a 4-year study period, no evidence of deterioration was observed and testing showed that there was no short-term breakdown of the linings. Water quality assessments were also conducted and showed that epoxy resin linings satisfied all test criteria (Warren, 1983).

Epoxy lining became widely used in the United Kingdom, but there were no national standards or guidelines published until WRc developed a work method in 1989 (WRc, 1989). Prior to 1989, individual water utilities controlled the application techniques, which led to a lack of consistency between lining applications. The following list presents common issues that often caused inconsistent and poor quality lining:

- A 4-hour cure time until approximately 1982
- No routine weight checks
- No metering devices for mix ratio control

- No use of spin-up time before lining
- No approved lining rigs, static mixers, and application heads
- No control over minimum cure times
- No pre- or post-lining CCTV surveys
- No limitation of lining in cold weather
- No QC procedures

APPLICATION TECHNOLOGIES

Application Procedures

Epoxy resin lining was the lining technique of choice in the United Kingdom, but was applied without national standards or guidance between 1979 and 1989; application was controlled by individual water utilities.

As application of epoxy lining increased, the *In Situ Epoxy Resin Lining – Operational Guidelines and Code of Practice* was published (WRc, 1989), followed by a second edition in 1997 (WRc, 1997). A third edition published in 2001 (WRc, 2001), titled *In Situ Epoxy Resin Lining—Operational Requirements and Code of Practice* (Code of Practice), is the current governing document. This document provides methods for applying epoxy lining, which provides the framework in which lining contractors should operate for in situ epoxy lining work. This document also acts as a basis for contractors' ISO 9000 Quality Assurance Manuals.

A summary of lining procedures is as follows (WRc, 1997):

- Preparatory work – A physical inspection of the pipe network should be conducted to determine the applicability of epoxy resin lining techniques. The section of pipe to be lined should be isolated.
- Mains cleaning and prelining inspections – Sections of pipe should be cleaned using a recommended method, such as power boring, drag scraping, pressure scraping, or pigging. A closed circuit television (CCTV) inspection should then be conducted to ensure that the pipe is free of debris and water, and is ready to receive the epoxy resin lining.
- Lining with epoxy resin – Ensure equipment is properly set up and that the lining rig consists of separate storage containers for resin and hardener components; a pumping system to supply epoxy materials to the application head; a monitoring system to check the mix ratio of resin and hardener; and a winch. The rig applies the lining by pumping the resin and hardener separately to the spinner head, where it is mixed immediately prior to application. The spinner head applies the lining uniformly to the interior of the pipe at a thickness of 1 mm.
- Curing and inspection – The lining must cure for a minimum of 16 hours and is subsequently inspected by conducting a CCTV survey. The lining thickness is also ensured by direct measurement at the ends of the lined pipe section.
- After proper disinfection, reconnect the lined section of pipe to the distribution system.

The development of epoxy resin lining techniques continues as it becomes more widely used in the United States. Over the past 30 years, materials research, conducted primarily in the United Kingdom, has led to the improvement of epoxy resins. QA/QC procedures also have become more stringent to ensure adequate lining techniques (Warren Associates, 1995). The Code of Practice published in the United Kingdom (WRc, 2001) is a primary reference in developing similar standards, recommendations, and guidelines in the United States for application of epoxy lining in potable water mains.

Quality Assurance and Quality Control (QA/QC)

QA/QC procedures are outlined in the Code of Practice for practices in the United Kingdom (WRc, 2001). In developing similar standards, recommendations, and guidelines in the United States, the following QA/QC procedures were considered to ensure linings are applied correctly to the interior surfaces of potable water mains (WRc, 2001):

- Routine weight checks
- Use of metering devices for mix ratio control
- Use of spin up time before lining
- Use of approved lining rigs, static mixers, and lining application heads
- Guaranteed control over minimum cure times
- Pre- and post-lining CCTV surveys
- Limitation of lining in cold weather

It should be noted that the failure to use or the lack of good quality CCTV equipment is one of the most critical omissions of QA/QC procedures. As a result, the actual quality of every epoxy-lined pipe length may be largely unknown.

In addition to QA/QC procedures during and immediately after the epoxy lining is applied, ongoing monitoring should be conducted to determine the effectiveness of the lining. Water quality monitoring will determine if water quality improves and hydraulic performance monitoring will ensure the capacity of the original pipe is restored.

Guidelines

AWWA is developing a standard for *Spray Applied In Situ Epoxy Lining of Water Pipelines* (AWWA, 2002). One basis for this standard is the Code of Practice (WRc, 2001), which provides similar guidance for lining practices in the United Kingdom. The draft AWWA standard is being developed to provide minimum requirements for design engineers and contractors to follow during the design, application, and inspection processes. It should be noted that AWWA standards do not provide all of the engineering and administrative detail contained within a specification. A *Manual of Practice for Rehabilitation of Water Mains, M-28,* has been published by AWWA (2001).

The AWWA draft standard and guidance document provide minimum requirements with respect to materials, design, application, and inspection of epoxy lining rehabilitations on pipe sections with a diameter 3 inches or larger. The following list summarizes the draft AWWA standard and Manual of Practice, A-28:

- Before an epoxy lining is applied:
  - The pipe must be structurally sound and free of corrosion and standing water.
  - Information about the service disruption must be distributed to water users.
  - A temporary water supply system must be provided.
  - The water used to clean the pipe must be properly disposed of.

  - The pipe must be inspected with a color CCTV system before and after epoxy lining application.
- An epoxy lining must be:
  - Machine-applied to straight and large radius bend pipe sections
  - Machine- or hand-applied to small radius bend sections of pipe
  - Applied in a one-coat process producing a coat thickness of 1 mm
  - Applied according to application equipment manufacturers' specifications
- The epoxy used must:
  - Be a two-component system consisting of two distinct colors
  - Be moisture-tolerant
  - Be able to be applied over a wide range of temperatures
  - Have strong adhesion characteristics in both dry and moist conditions
  - Produce a durable and smooth lining
  - Be certified according to the NSF 61 guidelines
- The epoxy lining should be inspected, after curing, for the following faults by a color CCTV system:
  - Uncured lining
  - Incomplete lining
  - Ringing
  - Slump
  - Water damage
  - Pinholes
- The lining should be evaluated for hydraulic performance using a performance criterion, such as the Hazen-Williams coefficient, and an internationally recognized certification program.

In addition to the AWWA standard and guidance documents, the epoxy resin materials used in lining processes must be certified under NSF 61. NSF 61 addresses indirect additives, products, and materials that come in contact with drinking water. Every material from the well or water intake to the discharge point is covered by NSF 61. The standard evaluates materials and products for contaminant leaching and certifies that the leached contaminant levels are within acceptable limits. The standard also evaluates a material's potential for microbial growth. When a material is certified under NSF 61, certification restrictions are imposed, such as maximum temperature or surface area to volume ratio (NSF, 2004).

To obtain NSF 61 certification, the NSF toxicology staff first evaluates the product or material submitted from the manufacturer. A factory audit is also performed during the certification process. During the audit chemical formulations and QA/QC records are inspected, and a sample of the material is obtained for further testing at an NSF laboratory. After the initial certification unannounced facility audits and material testing are performed on an annual basis (NSF, 2004).

Underwriters Laboratories, Inc. (UL) is an independent, not-for-profit product-safety testing and certification organization. UL's Drinking Water Certification Program is accredited by ANSI. UL's classification to NSF 61 helps clients meet local requirements and Code Authority requirements.

Faults in Epoxy Resin Lining Applications

Most lining faults cannot be tolerated because the effectiveness of the lining is compromised. Typical lining faults are summarized in Table 2.3 (AWWA, 2002; WRc, 2001).

Table 2.3
Summary of lining faults and remedial actions

Deficiency	Description	Prevention and remedial actions
Lining over debris	Caused by lining over corrosion deposits and/or loose debris in the pipe due to inadequate pipe cleaning that has not been identified by CCTV inspection. Premature failure of the lining is possible and it may decrease the expected hydraulic performance of the lined pipe.	If the fault is localized, it is best rectified by excavation and removal of the faulty pipe, which can be replaced. If the faulty section is extensive, it must be re-laid.
Blistering or bubbling	Round, raised sections of hardened epoxy caused where there is graphitization on cast iron pipe walls. It should be noted that not all instances of graphitization cause blisters.	If blisters are isolated, they may be cosmetic in nature and do not create a long-term weakening of the lining. In most instances, the integrity of the lining is not compromised and no remedial action is taken.
Uncured lining	Caused by a malfunction of the lining machine. Only sections of pipe lined with hardened, fully cured epoxy may be returned to service.	CCTV inspection will reveal any sections of uncured lining. The uncured material must be removed from the pipe and the section relined or replaced.
Incomplete or thin lining	Skips in the lining or sections of thin lining may be caused by machine malfunction, excessive friction along the lining hose as it is being winched back through the pipe, unsatisfactory design of the application head, or protruding obstacles in the main.	Depending on the length of the fault and the length of the section being lined, the contractor and the purchaser will decide on a course of action to remedy the situation. Short lengths can be replaced with new pipe and longer lengths should be re-cleaned and re-lined.

(continued)

Table 2.3 (Continued)

Deficiency	Description	Prevention and remedial actions
Ringing or ridged lining	Peaks and troughs in the lining caused by persistent jerking of the lining head in the main as a result of poor design or snagging of the hose on bends and fittings. Can also be caused by winch gear malfunction as it pulls the lining hose through the pipe.	The minimum thickness must be present in the troughs or over-lining will be required. Spot checks of thickness in troughs and peaks should be taken when camera inspection suggests significant ringing. In extreme cases, ridging peaks can reduce the desired friction coefficient and re-boring may be required.
Slump	Caused by applying an excessive thickness of epoxy to the pipe surface or overheating the components. Invert buildup up to 3/16 inches (5 mm) can be tolerated as long as the minimum thickness is apparent at the top of the pipe.	Spot checks of thickness in troughs and peaks should be taken when pipe end inspection suggests slump. In extreme cases, the volume of material can reduce the desired pipe diameter and re-boring may be required.
Water damage	Usually caused by lining through undetected standing water or water flow running along the invert of the pipe.	Depending on the length of the fault and the length of the section being lined, the contractor and purchaser together will decide on a course of action to remedy the situation. The extent of the problem should be accurately measured by careful CCTV inspection. Short lengths can be replaced with new pipe and longer lengths should be re-cleaned and re-lined.
Holidays	Pinhole or discontinuity in the lining that cannot be tolerated.	Holiday testing should be conducted to determine if there is a fault in the lining. If any pinholes are found, re-lining is required.

UTILITY CASE STUDIES/APPLICATION OF EPOXY LINING

Epoxy resin lining used for extending the life cycle of potable water mains has been widely applied in Europe for approximately 15 to 20 years, particularly in the United Kingdom. Research conducted in the United Kingdom and case studies examining in situ lining processes at various water utilities have provided a foundation for further research and application of epoxy lining throughout the world. A better understanding of epoxy resin materials, lining procedures, monitoring efforts, and QA/QC considerations has been achieved through previous in situ lining applications. For this literature review, epoxy lining applications at various utilities throughout the United Kingdom and North America have been reviewed and are summarized in the following subsections.

Yorkshire Water (Western Division)

Between 1982 and 1988 approximately 247 kilometers (km) of potable water pipes were lined with epoxy resin in the United Kingdom within the Western Division of Yorkshire Water (Warren Associates, 1997). According to the report (Warren Associates, 1997), the quality of epoxy resin linings in potable water pipes varied drastically during this time, which can be largely attributed to the fact that this method of rehabilitation was in its infancy and many of the mains lined were part of major trials. Published operational guidelines had not yet been released and QC methods were not as stringent as those implemented in later years. Investigations of pipes lined with epoxy during 1982-1988 ensued by gathering information, sampling pipes at various locations, and analyzing samples to determine the quality of epoxy lining. For each sample, the mix ratio of resin to hardener was determined and in many cases found to be incorrect, which led to the finding that approximately 50% of the linings were found to be in need of remedial work. There are many reasons that may have caused an incorrect mix ratio, some of which include (Warren Associates, 1997):

- Poorly trained contractor staff
- Poorly maintained epoxy-lining rigs
- Epoxy-lining rig was not set up with the correct mix ratio
- Lining procedures commenced before the correct mix ratio was achieved
- Weight checks of resin and hardener were not carried out and computer printouts were not produced to ensure the mix ratio was correct
- CCTV surveys were not used
- Incorrect cure time may have resulted in leaching of resin or hardener

North American Water Utilities

In North America, epoxy resin lining did not make an appearance until the mid-1990s. The first trial in the United States of in situ epoxy resin lining was in Chester, Pa. Chester Water Authority was chosen by AwwaRF as a demonstration site for the epoxy-lining process because of an excellent test site provided by Chester Water Authority (Conroy et al., 1993). The test main was a 6-inch, 80-year old unlined dead end cast iron water main.

The first phase of the project was an extensive literature review, followed by the selection of a test site. The study chose the rack and feed bore cleaning method and epoxy resin lining. Before the project commenced an extensive coordination effort was needed to approve the epoxy material and lining process because neither had any form of United States approval. After the epoxy lining had been applied it was concluded that the demonstration showed the application was of high technical quality. Inspection by CCTV showed excellent interior epoxy coverage (Conroy et al., 1993). A concern of epoxy leaching into the water supply was put to rest with a 30-day water quality monitoring program. The monitoring program concluded that leaching epoxy was well within water quality guidelines. The hydraulic performance gained with the epoxy-lining process was confirmed with flow test results showing an increase in the Hazen-Williams coefficient of roughness from 22 to 114 (Conroy et al., 1993).

The first projects involving the application of NSF 61-certified epoxy resin lining were carried out in Canada (AWWA, 2002). Application of an epoxy resin lining system as a nonstructural method to rehabilitate a pipe system is exemplified in the Region of Ottawa-Carleton. Epoxy-resin lining was selected because previous regional experience with this technology yielded an innovative, cost-effective solution to rehabilitating existing, structurally sound pipes (Marcuccio and Salvo, 2000).

In addition, several states and other Canadian provinces were involved with epoxy-lining projects during 1998 (Salvo, 1998). The states of Colorado, Indiana, Kansas, Minnesota, New York, and Pennsylvania undertook projects. In Pennsylvania, Bucks County Water and Sewer Authority (BCWSA) selected an area to apply epoxy resin lining techniques (Salvo, 1998). Ontario and Quebec also tested this technology. Table 2.4 summarizes epoxy lining projects in North America up to 1998.

Table 2.4
North American epoxy resin lining projects up to 1998

Client	Location	Length (ft)	Diameter (inches)	Contractor
Bucks County Water and Sewer	Pennsylvania	14,000	6	Nord Americain Construction
City of Pierrefonds	Quebec	6,600	6	Nord Americain Construction
City of Minneapolis	Minnesota	15,000	6 and 8	Cormar Contracting
Halton	Ontario	14,000	6	Cormar Contracting
Bucks County Water and Sewer	Pennsylvania	8,200	6	Philip Utilities
Montreal North	Quebec	8,000	6 and 8	AquaRehab
Peterborough	Ontario	30,000	6 and 8	Comar Contracting

DURABILITY TESTING

Laboratory Testing

Laboratory testing of epoxy resin lining is essential in determining the long-term durability and effectiveness of the material. Several laboratory studies were examined for this literature review. Generally, these articles discuss accelerated testing methods and ways to determine how long epoxy lining will be effective in protecting water mains.

Walker and Guan give a comparative study of the performance and application parameters of various NSF 61 approved lining systems that are presently specified and used in North America for the internal lining of municipal potable water mains (Walker and Guan, 1997). The following lining systems that were considered included: solvent-borne amine-based epoxy, 100% solids epoxy, 100% solids elastomeric polyurethane, 100% solids rigid polyurethane, and cement mortar lining. Laboratory testing was important to establish the materials' various performance attributes that form the basis in establishing correlation with field performance and long-term life expectancy.

Various parameters were tested to determine the quality of the protective lining and also the longevity of the lining systems. Table 2.5 summarizes the test parameters (Walker and Guan, 1997).

This study (Walker and Guan, 1997) concluded that among the epoxy resin lining systems, amine-epoxy corrosion protective systems offer good performance attributes including good chemical resistance, but are limited by other performance attributes such as poor flexibility and abrasion resistance. In addition, application difficulties common to epoxy systems are multiple coats, slow cure time, and an inability to cure at lower temperatures.

Another research effort (Canfer, 1990) discusses accelerating aging laboratory testing of epoxy resin lining. The purpose of accelerated aging tests is to determine the effects of age on the materials in a laboratory setting. For this research, a particular epoxy resin was tested for a duration of 8 years and 4 months under an accelerating factor of 12. This means that samples experience an exposure period equivalent to 100 years. Under these accelerated aging testing procedures, the longevity of the epoxy resin lining over a 100-year period was determined.

Test pieces consisted of quadrants cut from a section of pipe routinely lined in situ. The lining included deliberately added features, such as holidays in the epoxy and undermining of the lining around the cut edges of the test quadrant, for the purpose of encouraging corrosion of the cast iron substrate in specific areas. Test pieces were submersed in a medium consisting of de-ionized water and chloride as sodium chloride and sulfate as sodium sulfate. The test was carried out at 50 ºC.

At the conclusion of the study, the epoxy-lined samples were visually examined and found still to be adhering satisfactorily to the cast iron and exhibited no significant embrittlement when flexed by hand. Although the cast iron had corroded, the corrosion deposits had not lifted the lining away from the pipe surface. It was concluded that, when exposed to conditions that produce a corrosion of cast iron equivalent to 100 years under normal service conditions, the epoxy resin protected the metal surface that it covered.

Another study (Gray and Appleman, 2003) introduced the technique of electrochemical impedance spectroscopy (EIS), which can be used for evaluating coating permeability or barrier properties based on the electrical resistance of the coating. EIS involves measuring the lining's resistance to electrical current. In the laboratory, testing was conducted using EIS to determine coating performance for obtaining quantitative kinetic and mechanistic information on coating deterioration. Results showed a good relationship between coating impedance and coating field performance. Impedance is observed to decrease as a function of time of exposure to a corrosive environment. This decrease is observed to be related to the loss of barrier properties and the onset of under-film corrosion. EIS may be well suited as a field tool that can predict the remaining life of a protective coating.

In Situ Testing

Testing of in situ epoxy-lined pipes is essential in determining the quality of the lining procedures and the durability of the lining itself. To assess the long-term durability of epoxy resin lining, several studies have been published including one that describes the results of analyzing 82 pipe samples collected from 4 water utilities in the United Kingdom (Warren Associates, 1995). A test procedure was developed for exhumed pipe samples, which permits a quantitative assessment of the lining condition and quality, as well as an evaluation of its

Table 2.5
Test parameters for protective linings

Parameter	ASTM test	Definition and testing procedure
Abrasion resistance	ASTM D4060	Abrasion resistance testing represents the lining's ability to withstand the constant flow of liquid and any particulates. The Taber abrasion test was used, which involves rotating a sample under a specific weight against a grinding wheel for a defined number of revolutions. The samples were evaluated by measuring the weight of the sample before and after the test. The resulting weight loss indicated the comparative ability of the lining to resist abrasion and wear.
Adhesion	ASTM D4541	Adhesion of a protective lining system to the substrate is considered to be a good indicator of the lining's ability to resist corrosion and therefore, represents longevity of the lining. Generally, the better the adhesion, the longer the lining will last. The adhesion test method measures the "pull-off" strength of a lining system by determining the perpendicular force that the test material will withstand before either releasing from the steel surface or pulling apart cohesively.
Cathodic disbondment and salt spray resistance	ASTM G 95 and ASTM B117	Cathodic disbondment and salt-spray resistance both measure the undercutting resistance of a lining system. The two methods are compared because they both measure the same parameter. Cathodic methods were found to be more consistent in their ability to predict actual lining performance. Linings with better cathodic disbondment resistance have better corrosion resistance and greater longevity.
Chemical resistance	ASTM D714	Chemical resistance test methods monitor the effect of a chemical solution when the lining is applied to a metal coupon. The evaluation is completed by observing the panel for blisters and general appearance after immersion in the test solutions. While this test provides a "true-life" condition, the end result is that the lining can really only be recommended as being suitable for the length of service for which the sample was tested.
Flexibility	ASTM D552	Flexibility is an indicator of the lining's ability to withstand the cracking, disbonding, or other mechanical damage that can occur from handling and bending of the steel pipe, not only in the field, but also in the factory. Lined steel samples were bent to determine their ability to withstand bending before lining failure.

(continued)

Table 2.5 (Continued)

Parameter	ASTM test	Definition and testing procedure
Impact resistance	ASTM D2794	The impact resistance test method represents the lining's ability to withstand damage due to direct impact with another object. The test method consists of a fixed weight being dropped from varying heights to produce a point impact on the lining surface. The results are measured in terms of the energy required to rupture the lining and create a holiday or discontinuities. Energy is represented in terms of inch-pounds where 1 inch-pound is equivalent to dropping a 1-pound weight a distance of 1 inch onto a sample.
Water absorption	ASTM D570	Water absorption is the measure of the ability of a waterborne chemical or gas to penetrate the lining to the substrate. The test samples were immersed in potable water at 50 °C for 48 hours, and the weight of the samples before and after immersion in the water was noted. The results are reported in percentage of weight change, with the lower the resulting number, the better the lining is able to resist blistering and disbondment.

potential performance. The capacity of an applied lining to provide protection to distribution systems over an acceptable time scale is dependent on a number of factors relating either directly to the resin formulation or contractor performance (Warren Associates, 1995; Warren and DeRosa, 1996):

- The physical and chemical properties of the epoxy resin (at optimum mix ratio)
- The sensitivity of the epoxy resin to mix ratios
- The thoroughness of pipe internal surface preparation prior to lining
- The sensitivity of the epoxy resin to application conditions

FUTURE RESEARCH

Further research efforts are needed to determine how long a pipe's life-cycle is extended when epoxy lining is applied. A testing procedure or protocol that helps to determine the useful life of a lined pipe is necessary. In addition, monitoring techniques should be reviewed and applied so that the effectiveness of epoxy lining can be measured.

In addition to epoxy resin lining, other materials have recently been considered for in situ pipe lining, including rapid setting polyurethanes, polyureas, and other isocyanate-cured materials. Polyisocyanate-based composites have had widespread use as protective coatings in various industries ever since a process for making polyurethanes and polyureas was developed in 1937 (Warren Associates, 2000). An advantage of using these materials is their ability to cure rapidly. Epoxy resin has a lengthy curing time, which causes significant water disruption while the pipe rehabilitation procedure is being completed.

Although polymeric linings are not currently used in North America, the United Kingdom has already generated operational guidelines for the application of polymeric pipe quantitative assessment of the lining condition and quality, as well as an evaluation of its linings (Warren Associates, 2000). Polymeric linings have a very short curing time (a few hours), which allows water utilities to rehabilitate a pipe in less than one business day. Polymeric linings are formulated from a solvent-free two-component system. The application process of polymeric linings is very similar to the epoxy-lining application process (Warren Associates, 2000).

In a comparative study among 100% solids epoxies, 100% solids elastomeric polyurethane, and 100% solids rigid polyurethane linings (Walker and Guan, 1997), it was determined that the 100% solids rigid polyurethane system was the best performing liner. Rigid polyurethane was determined to be the best system due to its high tensile adhesion, low undercutting, and excellent abrasion and impact resistance. These attributes make 100% rigid polyurethane a high performance corrosion-protection system that can be used in cases of aggressive environments, repeated hydration and dehydration, high flow velocities, and possible pipe strains (Walker and Guan, 1997). Also, 100% solids elastomeric polyurethane was determined to provide good abrasion resistance and flexibility, but poor adhesion to the steel substance and an inability to resist undercutting (Walker and Guan, 1997). The difference between rigid and elastomeric polyurethanes is the types of isocyanate and polyol used in resin formulation. Rigid polyurethanes have a highly cross-linked chemical structure, which creates a hard, dense protective membrane, whereas elastomeric polyurethanes have a more linear chemical structure, which results in high flexibility and poor tensile adhesion (Walker and Guan, 1997).

CHAPTER 3
SURVEY OF CURRENT PRACTICES

PRELIMINARY SURVEY OF EPOXY LINING

A preliminary survey of water utilities was conducted to determine the extent to which cleaning and epoxy lining have been performed in their systems. The purpose of this survey was to develop a matrix of the types of epoxy materials used and the years in which they were applied. A pipe sampling program was developed from this matrix with the objective of collecting samples of epoxy-lined pipes covering a wide range of epoxy materials and ages. In order to identify water utilities having experience with water main lining with various epoxy materials, the following information was requested from water utilities in North America and the United Kingdom.

- Year of installation of lining
- Pipe diameter
- Length of pipe lined
- Epoxy manufacturer
- Epoxy product
- Contractor
- Cleaning method used
- Year pipe was installed
- Pipe material

The results of the preliminary survey are summarized in Table 3.1. It is apparent from the responses that the practice of epoxy lining is more prevalent in the United Kingdom than in North America.

Table 3.1
Preliminary questionnaire results

Water utility	Amount of epoxy-lined pipe (feet)	Installation of epoxy lining		Epoxy type	Range of pipe diameters	Pipe materials lined
		Start year	End year			
Denver Water	24,800	1998	2001	Type 1	6-inch to 12-inch	Cast iron
Halifax Regional Water Commission	43,600	2000	2003	NA	6-inch to 15-inch	NA
Chester Water Authority	470	1993	1993	Type 1	6-inch	Cast iron
Bucks County Water and Sewer Authority	29,500	1997	1999	Type 2	6-inch to 8-inch	NA
Yorkshire Water	9,137,600	1989	2002	NA	3-inch to 4-inch	NA
Thames Water	17,371,200	1991	2002	Type 1 and 2	3-inch to 12-inch	Cast iron and ductile iron

NA = Not available

WATER UTILITY QUESTIONNAIRE

More detailed questionnaires were sent to 11 water utilities who indicated that they used epoxy to line water mains. Replies from the following eight water systems in the United Kingdom were received regarding the use of epoxy to line water mains.

- Guernsey Water
- Severn Trent Water
- South East Water
- South West Water
- Thames Water
- Veolia Water
- Wessex Water
- Yorkshire Water

The water utility questionnaire requested information about the reasons why epoxy liners were used, the performance of in-place liners, and the installation and QC procedures used by the utility. A copy of the questionnaire sent to water utilities is included in Appendix A.

The water utilities responding to the questionnaire said that the most frequent reason for epoxy-lining a pipe (Question 3) was water quality complaints. All reported improvement in water quality as a result of epoxy lining. Other reasons cited for using epoxy lining were: (1) epoxy lining is a fast, effective, and cost-effective method for rehabilitation of water mains; and (2) it provides the best whole-life value (i.e., lowest life-cycle cost). No respondents mentioned hydraulics (C-factor) as the reason for epoxy lining their pipes. The performance of the epoxy liners was categorized as "OK so far" by half the respondents, and "very good" by two water utilities. One respondent noted that early applications caused water quality problems and that "vastly improved epoxy materials and cleaning procedures" have corrected these problems. Another respondent said that the performance of epoxy liners is very good now after the use of CCTV inspection methods.

When asked about the occurrence of leaks and breaks after epoxy lining (Questions 5 and 6), half the respondents said that there was no change in the frequency of leaks and breaks. One utility said that the rate of leaks and breaks increased by 15% after epoxy lining and another said that the rates were lower.

All eight utilities said that water quality improved after epoxy lining (Question 7). Improved chlorine residual was specifically mentioned by three respondents and lower turbidity was mentioned by one utility. All respondents said that hydraulics were improved after epoxy lining (Question 8), although improvements in Hazen Williams C-factor or water quality k-factor were not measured or quantified in most cases. It was emphasized that improved hydraulics was "not the main driver" for using epoxy liners.

Three respondents had inspected the epoxy liners after the pipes were put back into service (Question 9). The only problem cited was disbondment when epoxy lining was placed over cement mortar lining. Five utilities collected pipe coupons after epoxy lining (Question 10), but none performed tests on the coupons (Question 10).

All utilities said that installation reports are available (Question 11) along with written installation procedures (Question 12). These are required by the Code of Practice (WRc, 2001) for installation of epoxy resin liners in the United Kingdom. The Code of Practice also requires documentation of the pipe condition prior to lining (Question 13). CCTV is the primary QC tool used to ensure the quality of the pipe cleaning prior to epoxy lining (Question 14). The use of CCTV after cleaning is required by the Code of Practice (Question 15).

The quality of the applied epoxy liner is ensured by post-application CCTV inspections as well as rig printouts, rig calibration, and rig audits. The QC procedures for verifying the quality of the epoxy liners after installation are specified in the Code of Practice (Questions 17 and 19). QC procedures include CCTV inspections, review of lining records, and pipe samples. All respondents said that they perform CCTV inspections before and after applying the epoxy liner (Question 22).

The United Kingdom water utilities replied that they expected the useful life of an epoxy liner to be 30 to 35 years (Question 20). Two respondents said that the useful life would be 50 years or greater, and one utility said that they expected the useful life of an epoxy liner to be "similar to a new (pipe) installation." All respondents said that they were satisfied with the performance of the epoxy liner (Question 21).

The water utilities in the United Kingdom said that the water mains could be put back into service a minimum of 16 hours after lining (Question 23).

EPOXY MANUFACTURER QUESTIONNAIRE

Questionnaires were sent to epoxy manufacturers requesting information about the physical properties of their epoxy products, including Material Safety Data Sheets (MSDSs).

These questionnaires were sent to five epoxy manufacturers in order to obtain material properties for various epoxy products. Responses were received from four epoxy manufacturers (E. Wood, ELC, Mercol, and Raven Lining Systems) for a total of seven epoxy formulations and one polyurethane formulation. A copy of the questionnaire sent to the epoxy manufacturers is included in Appendix B.

The properties of the epoxy formulations provided by the manufacturers show significant differences in material properties, summarized as follows:

- Cure time – between 5 hours and 16 hours
- Adhesion – between 1,000 pounds per square inch (psi) to 2,500 psi
- Barcol hardness between 87 and 120

Table 3.2 summarizes the material properties reported by epoxy manufacturers.

In general, the physical properties of the epoxies listed in Table 3.2 are quite similar to each other.

Table 3.2
Summary of epoxy material properties from manufacturers' questionnaires

Parameter	Copon Hycote 162PWX	Copon Hycote 169 (Rapid setting polymeric lining)	AquataPoxy A-61	AquataPoxy A-6	ELC 173/90	ELC 257/91	Geopox GX014
Solids content of resin	100%	100%	100%	100%	100%	100%	100%
Benzyl alcohol content	nil		0%		0%	0%	0%
Mix ratio (resin:hardener) by weight	100:60	100:38.8			100:40	100:32	100:42
Mix ratio (resin:hardener) by volume	2:1	2.5:1	3:1	1:1			2:47.1
Recommended film thickness, mm	1 minimum	1 minimum	0.25-12.7	0.2-0.66	1	0.2	1
Recommended film thickness, mils			10-500 mils	8-26 mils			
Recommended pipe temperature for application	>3 °C	>3 °C	>40 °F	>40 °F	>3 °C	>3 °C	>3 °C
Pre-heat temperature	40-50 °C	25-35 °C	110-120 °F	140-190 °F	15-30 °C	20-45 °C	15-30 °C
Hardening time (hours)				72			
Cure time (hours)	16	0.5	5	72	16	16	16
Return to service time (hours)	16	0.5	5	72	16	16	16
Full mechanical service	7 days	24 hours					
Viscosity of resin at application temperature	n/a	n/a	n/a		44,000 Mpa @20°C	46,000 Mpa @20°C	30-40 Poise @ 20°C
Viscosity of hardener at application temperature	n/a	n/a	n/a		6,500 Mpa @20°C	28,000 Mpa @20°C	50-60 Poise @ 20°C
Hardness	120 Barcol	120 Barcol	88 Shore D	87 Shore D	88 Barcol	86.7 Barcol	
Adhesion, pull-off	70 kg/sq. cm	70 kg/sq. cm	>1,550 psi	>2,500 psi	n/a	n/a	
Impact resistance	2.6 Joules	2 Joules			n/a	n/a	
Tensile strength							26 Mpa
Tensile modulus							2,947 Mpa
Flexibility	0.6%	0.5%		n/a			
Flexural modulus					2,568 MPa	1,844 MPa	
Flexural strength					79 MPa	53 MPa	38.41 MPa
Flexural strain at failure					2.8%	2.5%	
Shrinkage to 20 °C							0-5%
Wear resistance					0.1976 Tabor	0.1724 Tabor	0.181775 Tabor
Water absorption			3.59%	2.22%			
Length of time product has been in use, years	10	4	1	17	14	10	18

CHAPTER 4
LABORATORY ANALYSIS OF HISTORICAL EPOXY LININGS

INTRODUCTION

One of the most important aspects of this research was to identify and quantify various parameters of historical epoxy lining installations and to correlate them with historical performance of the liner. Specific historical epoxy lining projects were identified after consultation with five water utilities in the United Kingdom and North America for pipe sample collection. Selection of pipes was made to cover various epoxy products and various dates of installation that are in use in water main lining.

COLLECTION OF SAMPLES

A total of 17 different samples of epoxy-lined water mains were obtained from Yorkshire Water, U.K.; Denver Water, Colo.; Halifax Regional Water Commission (HRWC), Cana.; Chester Water Authority, Pa.; and Bucks County Water and Sewer Authority (BCWSA), Pa. Each sample consisted of intact pipe cut from an operating water line. Physical observations of the condition of the linings and the pipe samples and other information were noted. The samples represented a range of epoxy products and service times. Installation dates ranged from 1979 to 2003. Two of the samples (sample #101 from Denver Water and sample #107 from BCWSA) consisted of pipes with a cement liner, over which the epoxy had been applied. With the exception of Chester Water Authority and Denver Water, all of the samples were packed dry and had dried out prior to receipt at the project laboratory, KTA-Tator Laboratory, Cana. Newly applied samples were not available for comparison because none of the lining rigs were active at the time.

A total of 10 pipe samples were taken across the Yorkshire Water service area covering the first linings of 1979 to a sample from a pipe lined in 2000. The 10 samples also covered the 5 different epoxy resin materials that have been applied in the Yorkshire service area. The pipe samples were each 1.2 meters (m) (4 ft) long and they were split into two equal lengths for shipping to the KTA-Tator Laboratory, Cana. for testing. All pipe samples were 3-inch and 4-inch diameter cast iron because this reflected the bulk of the pipe sizes for which epoxy lining has been undertaken by Yorkshire Water. The samples were taken by hydraulic pipe cutters in order to avoid heat damage to the lining. BCWSA and Chester Water Authority used snap cutters to remove pipe samples from the ground. At the same time that the pipe sample was taken a record was made of the soil type and condition.

Historical information for all of the pipe samples was recorded, which included the date that the lining was performed, the contractor who applied the lining, the code of practice that was in force at the time, and the epoxy resin material used.

Before the pipe samples were shipped for testing, and after receipt by KTA-Tator, they were photographed. The photographs are shown in Appendix C. The photographs showed that there was a wide variation in appearance, which meant that a wide range of data would be possible from the testing. Some of the linings were still in excellent condition after a number of years in service and were still providing a good barrier between the water and the cast iron pipe. Some of the linings were showing evidence of iron coming through the lining, and one of the linings appeared to have failed completely.

A total of eight epoxy materials were tested, of which six were of Type 1 and two were of Type 2. The list of epoxy materials is shown in Table 4.1.

Table 4.1
List of various epoxy lining materials tested

Epoxy Lining Materials Tested	
Huntington Industrial	Type 1
Geopox 012	Type 1
Geopox 014	Type 1
Nitoline WP	Type 1
Nitoline WP2	Type 1
ELC 173/90	Type 1
ELC 257/91	Type 2
Hycote 169 PWX	Type 2

LINING TESTING PROCEDURE

Upon arrival at the laboratory the samples were systematically unpacked, along with related identification materials and service information. The samples were photographed to document the appearance of the lining and pipe as received. The samples were given a preliminary inspection for the condition of the lining, blistering, corrosion tubercles, and any other observations.

After the initial examination and documentation of observations, one of the two pieces of pipe for each lining was cut in half along its length to facilitate detailed examination. The pipe was cold cut to avoid damage to the lining. The linings were fully dried out prior to examination so that a fair comparison could be obtained between them. The following detailed analyses were conducted on the cut sample:

- General observations of condition and application
- Blistering
- Dry film thickness (DFT)
- Holiday test
- Knife adhesion
- Corrosion of the cast iron under the lining and at defects
- EIS measurements
- Microscopy

Details of testing procedures are given in Appendix D.

SUMMARY OF OBSERVATIONS OF PIPE SAMPLES

The physical observations for the 17 pipe samples are outlined in Table 4.2. Photographs of all of the field samples are shown in Appendix C. The findings are summarized in the following subsections.

Table 4.2
Physical observations and DFT of field samples

Sample no.	Origin	Lining type	Diameter (inches)/ material	Lining installation date	DFT (mils)	General condition	Abrasion
1	Yorkshire Water UK	Type 1	4" Cast Iron	1984	19	Poor	No
3	Yorkshire Water UK	Type 1	3" Cast Iron	1991	32	Excellent	No
4	Yorkshire Water UK	Type 1	4" Spun Iron	1996	47	Excellent	Some dulling of the surface, possibly from red corrosion product deposition
5	Yorkshire Water UK	Type 1	3" Cast Iron	1992	37	Excellent	None
6	Yorkshire Water UK	Type 1	4" Cast Iron	2000	47	Excellent	None

(continued)

Table 4.2 (Continued)

Sample no.	Origin	Color change	Gloss	Blisters (ASTM D714), volcanoes	Holidays, (no. per square foot)	Coating	Ringing, pooling, other application defects
1	Yorkshire Water UK	No reference	Glossy	Many; 1/4 to 1 inch in size; blisters are only 1/2 to 1 inch apart; lower density along one side of the pipe; rust spots on many blisters	Continuous	Hunting Industrial	Continuous, 1/8 to 1/4 inch rings, some minor runs but does not look like pooling
3	Yorkshire Water UK	Very slight	Little to none	None; somewhat irregular surface	0.8	Geopox 014	Very little ringing with only one large ring (0.5-inch band next to thin spot); pooled area at one end, 3/16 inch thick, 2 inches wide
4	Yorkshire Water UK	Slight bleaching	Slight	None on the sample examined. A few isolated blisters 1/4 inch in size on other pipe section	0	Geopox 014	Minimal to none, slight irregularity in surface smoothness
5	Yorkshire Water UK	Very slight bleaching	High	A few isolated blisters and open volcanoes, less than 1/4 inch in size	5	Nitoline WP2	Minimal to none, but finish is somewhat rough with splatters
6	Yorkshire Water UK	Very slight bleaching	Glossy	A few open volcanoes, including a longitudinal cluster at the crown, less than 1/8 inch	14	Nitoline WP	None. 6-inch section of runs and sags from crown, but oriented in the wrong direction for running down from the marked crown. Some globules of coating and volcanoes. Possible "pool" of coating is actually covering a band of bituminous coating not removed

(continued)

Table 4.2 (Continued)

Sample no.	Origin	Rusting through coating ASTM D610	Color	Parallel scribe adhesion	Steel around holidays and blisters	Steel under intact coating
1	Yorkshire Water UK	5G at the center of most blisters; 8P at the numerous holidays	Pale tan	Rating 1 to 3. Some areas of coating offer resistance to lifting with the knife (i.e., rating 8).	Some blisters were accompanied by visible areas of minor metal corrosion.	Rusted but only minor apparent mass loss; a dust of corrosion products could be brushed away. Irregular shapes on the surface are probably not from metal loss but were the original morphology of the pipe surface.
3	Yorkshire Water UK	None, including at a holiday; some surface rust speckles	Red - slight bleaching, matte	Rating 2 to 4. Coating can be pried from the surface with a utility knife, but with difficulty.	Coating cut and lifted around one holiday.	Corroded, but with little mass loss since the coating was applied. Longitudinal lines, tiny nodules, and geometric cuts all have sharp edges with no build-up of corrosion products.
4	Yorkshire Water UK	Rusting at one holiday; deposit of rusty scale from 10:30 to 1:30 position	Red with bleaching, matte.	Rating 2 to 3. Coating is hard to remove, however, and extra strong utility knife is required to clear the coating from the surface for metal examination.	Not applicable. No deterioration of mortar under holidays.	Some rusting with black corrosion products; looks like the pipe was in very good condition before the coating was applied.
5	Yorkshire Water UK	Rust at the holidays and other defects. Some rust products deposited over the surface at bottom portion of pipe.	Dark blue, glossy	Rating 0 to 1. Coating can be lifted in large chips. Some areas have slightly better adhesion.	No noticeable change in corrosion around two holidays Large rust stain under a third holiday.	Rusted, red and black scale. The corrosion is relatively uniform, but generous amounts of red and black corrosion products are present. The red corrosion products in particular imply more active corrosion.
6	Yorkshire Water UK	None, except at volcanoes	Dark blue, glossy	Rating 2 to 3. In some areas a strong prying action is required to remove the coating; in others the coating is very loose.	No noticeable difference at holidays.	Rusted, mostly black corrosion products.

(continued)

Table 4.2 (Continued)

Sample no.	Origin	Lining type	Diameter (inches)/ material	Lining installation date	DFT (mils)	General condition	Abrasion
8	Yorkshire Water UK	Type 1	3" Cast Iron	1995	44	Excellent	None
9	Yorkshire Water UK	Type 1	4" Cast Iron	1995	32	Excellent	None
10	Yorkshire Water UK	Type 2	4" Cast Iron	1995	62	Medium	None
11	Yorkshire Water UK	Type 2	3" Cast Iron	1995	56	Excellent	None
12	Yorkshire Water UK	Type 1	4" Cast Iron	1979	38	Poor	Yes, or significant deterioration
101	Denver Water	Type 1	6" Cast Iron	2001	54	Excellent	None or minor except for what looks like a scrape about 3/4 inch in size on the smooth side.

(continued)

Table 4.2 (Continued)

Sample no.	Origin	Color change	Gloss	Blisters (ASTM D714), volcanoes	Holidays, (no. per square foot)	Coating	Ringing, pooling, other application defects
8	Yorkshire Water UK	No reference; pale yellow, no change with DFT	Glossy	None obvious; occasional pinholes	6	ELC 173/90	Some large bands, thick alternating thin spots; very irregular application with an irregular finish and folds
9	Yorkshire Water UK	No reference; pale yellow, no change with DFT	Glossy but with porous areas probably due to a thin film over a rough pipe surface	Clusters of volcanoes at 6 o'clock, less than 1/4 inch; some holidays	26	ELC 173/90	Some runs and sags, with pooling at 6 o'clock
10	Yorkshire Water UK	Bleaching	Matte finish, less than original gloss	Isolated volcanoes and holidays; also some small mounds of coating (sputter)	Continuous	ELC 257/91	Severe, with wide bands (1 inch) of alternating thick and thin coating (4-inch repeat interval); thin and thick areas are 20 and 50 mils thick, respectively.
11	Yorkshire Water UK	Bleaching	Semi-gloss	None; a few isolated holidays and small volcanoes	0	ELC 257/91	None. Some minor sags
12	Yorkshire Water UK	Bleaching, and saturation with corrosion products	None	A number of large tubercules clustered toward the 6 o'clock position, over about 100 degrees of arc; tubercules are 1/2 to 2 inches in size; unclear if the red coating is gone from the tubercules.	Continuous	Geopox 012	Minor in one sample.
101	Denver Water	Slight bleaching	Semi-gloss	None. Isolated holidays; clusters of very small rust spots on the surface in a longitudinal strip of coating at 6 o'clock.	1	Geopox 014	Minor, 1- to 2-inch bands at 3 to 9 o'clock; drips at remaining crown area.

(continued)

Table 4.2 (Continued)

Sample no.	Origin	Rusting through coating ASTM D610	Color	Parallel scribe adhesion	Steel around holidays and blisters	Steel under intact coating
8	Yorkshire Water UK	None, except some along a very rough application area	Yellow, glossy	Rating 0. Coating readily comes off in large pieces. Some of the roughness in finish is due to underlying roughness. Crown marking seems accurate based on pooling of coating pattern.	No noticeable difference at holidays	Rusted, mostly black corrosion products
9	Yorkshire Water UK	At a few holidays and volcanoes	Yellow, glossy	Rating of 2 to 3. Adhesion seems good, but may be due to low DFT and weakness of the film. Only small chips can be pried from the surface.	No noticeable change; hard to pry away a piece of coating	Rusted, mostly black corrosion products; no difference at thin areas that failed the holiday test.
10	Yorkshire Water UK	At a few holidays and volcanoes; rust speckles on surface of coating	Pale yellow, matte	Rating 0 to 1. Coating can be lifted in large chips. Some areas have slightly better adhesion.	Steel is in generally good condition with a film of corrosion products.	Rusted, with mostly black corrosion products; no apparent effect from thin and thick areas of rings.
11	Yorkshire Water UK	At a few holidays and volcanoes; coating surface has rust speckles	Pale yellow, glossy	Rating 0. Coating readily comes off in large pieces.	Not applicable. No deterioration of mortar under holidays.	Rusted, with mostly black corrosion products; pipe was in good condition when coated.
12	Yorkshire Water UK	Yes, coating looks porous, although the coating is red similar to the rust. A deposit of red corrosion products over most of the coating.	Red, considerably bleached, rough surface	Rating from 0 to 3. Coating melted when cut with Dremel tool. When the parallel cuts were made with a utility knife, the coating broke away in large pieces.	The voluminous corrosion products at the tubercule suggest significant corrosion. However, visible mass loss is small.	Rusted, with mostly black corrosion products; no significant mass loss under turbercule.
101	Denver Water	None. Coating was applied over cement mortar, varying from 1/16 to 3/16 inch in thickness.	Red, glossy, some bleaching	Rating 0. Entire coating can be lifted in one piece from cement mortar.	Not applicable. No deterioration of mortar under holidays.	Coating was applied over mortar. Good physical contact between epoxy coating and cement liner.

(continued)

Table 4.2 (Continued)

Sample no.	Origin	Lining type	Diameter (inches)/ material	Lining installation date	DFT (mils)	General condition	Abrasion
102	Denver Water	Type 1	6" Cast Iron	1998	63	Excellent	None or minor.
103	Denver Water	Type 1	6" Cast Iron	1999	53	Fair	None or minor
104	Chester Water Authority, Pa.	Type 1	6" Cast Iron	1993 (1912 line)	47	Fair	Abrasion possible, as coating gloss is gone.
105	Halifax Regional Water Commission	Type 1	6" Cast Iron	2002 or 2000	52	Good, although there is a heavy film of corrosion products over the surface.	Slight, some abrasion marks, which look mechanical.
106	Halifax Regional Water Commission	Type 2	12" Cast Iron	2003	49	Poor	None
107	Bucks County Water and Sewer	Type 2	7.5" Cast Iron	2000	49	Good	None to slight

(continued)

Table 4.2 (Continued)

Sample no.	Origin	Color change	Gloss	Blisters (ASTM D714), volcanoes	Holidays, (no. per square foot)	Coating	Ringing, pooling, other application defects
102	Denver Water	Slight bleaching	Semi-gloss	None; a number of craters 1/8 inch in size, one with no coating at bottom of crater.	0.3	Geopox 014	Moderate banding with about 1-inch separation; some large "dimples" where the film may be thinner.
103	Denver Water	Slight bleaching	Semi-gloss	Two to three large blisters (1 to 1.5 inches) in each pipe section. Numerous holidays and shallow volcanoes, some with little coating at the bottom (1/8 inch). Some blobs of coating (1/8 inch).	25	Geopox 014	Moderate to severe banding, separated by several inches. Bands about 2 inches wide, with 3/4 to 1 inch separation. Banding is more prominent on half of the sample.
104	Chester Water Authority, Pa.	Extensive surface bleaching; surface is pale pinkish yellow.	None, matte finish	Several irregularly shaped, swollen areas of coating that look like blisters. However, they are covering rough areas on the pipe; a number of small holidays and volcanoes (1/8 inch).	6	Geopox 014	Moderate, 1/2 inch spacing over most of the sample
105	Halifax Regional Water Commission	Coating is medium grey, possible minor surface bleaching.	Glossy	No blisters. Two large (0.5 to 0.75 inch) rounded thin areas.	1	Hunting Industrial	Minimal; application looks quite even except for two thin areas.
106	Halifax Regional Water Commission	Moderate, from light grey to military olive green on the surface.	Glossy	Cluster of 5 blisters, each about 1 to 1.5 inches in size; numerous volcanoes (1/8 inch) with pinholes; line of "blisters" is actually spray dribbles.	7	Hycote 162PWX	Slight ringing, irregular pattern of application with some globs (1/8 inch) of coating, no pooling; a long line of coating dribbles (0.4 inch wide) along the length of the pipe
107	Bucks County Water and Sewer	Slight darkening	Semi-gloss	None. Isolated pinholes. No significant other defects.	3	ELC 257/91	Minor ringing on one side of the sample. No pooling.

(continued)

Table 4.2 (Continued)

Sample no.	Origin	Rusting through coating ASTM D610	Color	Parallel scribe adhesion	Steel around holidays and blisters	Steel under intact coating
102	Denver Water	None. Some rust speckles on the surface; some surface discoloration with a pattern that looks like a 3-inch wide superficial scrape along the length.	Red, glossy, some bleaching	Rating 2 to 3. Coating is hard and can be pried from the surface by slipping a utility knife under it. Large pieces break away.	Shiny patch around one holiday, but no noticeable increase in corrosion.	Rusted, mostly black corrosion products with little mass loss.
103	Denver Water	Rust speckles on the surface of the coating. Only occasional rusting at holidays and volcanoes.	Red, matte, varying degrees of bleaching	Rating 0 to 2. Coating can be pried from the surface with a strong knife or screw driver. Where the coating is strong, large chips can be removed. Where the coating is thin, only small chips can be removed.	No significant mass loss associated with holidays or large blisters. Boundary of the blister is faintly marked by a ring, with little noticeable metal loss.	Rusted, mostly black corrosion products. Some patches of yellow and red corrosion products were observed under pieces of "intact" coating lifted from the surface. Possibly the corrosion products were present prior to coating application.
104	Chester Water Authority, Pa.	Rusting at some holidays and volcanoes; rust speckles on surface	Red, with surface bleaching to light tan; matte	Rating 0. Some areas appear to have better adhesion, but are only thinner DFTs. Large (2 x 4 inch) pieces to small pieces (1/4 inch size) can be removed with a screwdriver and hammer, or large utility knife.	No noticeable corrosion around holidays.	Rusted, mostly black corrosion products; surface cleaning tool seemed to mark the surface easily (i.e., from graphitized steel?). Small amount of yellow/red corrosion products associated with the rough area on the surface (that looked like a blister).
105	Halifax Regional Water Commission	Heavy deposit of corrosion products over the surface, including a 2-inch wide band running the length of the sample	Medium grey	Rating is 0. Coating can be removed in large chips with a knife or screw driver and hammer.	No noticeable difference. No visible increase in corrosion products associated with any of the defects.	Much of the surface is graphitized and soft. A thin film of red rust products covers most of the surface, which looks more corroded than most of the other samples.
106	Halifax Regional Water Commission	None. Deposit of rust on the surface of the coating suggesting upstream corrosion.	Army green surface, light grey inside coating film	Rating of 0. The coating has shrunk away from the side of the pipe next to the cut area. The coating can readily be removed by prying.	No noticeable difference.	The steel under the blisters shows a definite corrosion pattern related to the blisters; possibly some type of contamination was present when the coating was applied, which enabled the corrosion to proceed.
107	Bucks County Water and Sewer	None. Coating is applied over cement mortar	Olive green	Rating 0 to 1. Large pieces can be pried off the surface.	Not applicable, as coating is applied over mortar.	Coating applied over mortar.

General Condition

The epoxy linings were generally in good to excellent condition. The exceptions were samples #1, #12, and #106, which were in poor condition. Sample #1 had a high density of large blisters with rusting and considerable physical deterioration. Sample #12 had a number of large tubercles with rusting and severe physical deterioration; this epoxy is no longer in use. This sample lining was constructed in 1979 without any proper protocols. Sample #106 had a few blisters and a large number of holidays with rusting; the epoxy lining, where intact, did not show significant physical deterioration.

Sample #104 was in generally good condition, except for significant bleaching and loss of gloss.

Unlike any of the other epoxy linings, sample #12 melted when subjected to the Dremel tool cutting wheel by which the cuts were made for the parallel scribe adhesion analysis. Possibly there had been a problem with this lining application in mixing or curing.

Abrasion, Erosion, Surface Bleaching, and Color Change

There was little, if any, abrasion or erosion of the epoxy lining surfaces based on their retention of gloss. The exceptions were samples #104 and #12, which showed a complete loss of gloss compared to other samples and to laboratory preparations. It was uncertain, however, if the loss of gloss was related to erosion or occurred for other reasons.

Most of the linings showed minor color changes and/or bleaching, which is normal for epoxies in water service. The color change was limited to the surface of the lining, which is in contact with the potable water. The exceptions were samples #104 and #10. Sample #104 showed significantly more bleaching than any of the other similar material samples. Sample #10 showed bleaching, which varied with the ringing pattern, where the thin areas of the coating bleached significantly more than the thick areas.

Application Irregularities

Various irregularities were observed in the epoxy linings, which were related to application of the epoxy lining. These were the following:

- Ringing: Ringing consists of alternating bands of thick and thin epoxy along the length of the pipe. The “rings” or “bands” were formed during application of the lining. The spinning head that was applying the coating was drawn through the line in a jerky, rhythmic manner instead of a smooth motion.
  The samples showed varying degrees of ringing, from none to severe. Sample #10 had the most severe ringing, consisting of 4-inch wide rings with maximum and minimum DFTs of 20 and 50 mils, respectively. Others with significant ringing were in sample #8 (with a few large rings), sample #102 (thick 1-inch rings separated by a narrow thin ring), and sample #103 (moderate to severe 2-inch rings).
- Pooling: Pooling consists of excessive coating thickness along the bottom of the pipe when excess liquid coating flows and collects at the bottom of the pipe, prior to curing. Only two samples showed any significant amount of pooling: sample #3 ($^3/_{16}$-inch thick coating) and sample #9 (pooling). Sample #6 looked like it had a large pool of coating,

but detailed examination showed that the epoxy lining was covering a strip of coal tar coating that had not been removed from the underlying metal.

- Runs, sags, dribbles: Runs and sags consist of liquid coating running in large drip-like runs prior to curing of the lining. Dribbles are large drops of coating resulting from improper spin head operation. Minor instances of all of these irregularities were observed on many of the samples.
- Rippling: Rippling consists of an irregular surface that looks like water ripples. This type of irregularity can be caused by the spinning head (excessive air directed onto the lining) or flow of the lining material due to low viscosity if it cannot "hang" on the surface as a smooth film. Sample #101 had particularly severe rippling over half of its surface.

Although these irregularities are undesirable, they may not necessarily compromise performance, provided that the epoxy lining DFT is at least 40 mils and the coating is properly cured with good barrier properties. Excessive DFT, however, can lead to high residual stress within the lining from higher shrinkage during curing. This could lead to the lining pulling away from the pipe wall and cracking as the line is taken out of service and allowed to dry out. The irregular surfaces could also promote accumulation and adhesion of deposits from the water over many years of service.

Blisters

There were no blisters on 12 of the 17 samples. Upon initial inspection, structures were observed on some of the 12 samples that appeared to be blisters (samples #103, #10, and #104). Close examination showed that the "blisters" were either thickened globules of epoxy from irregularities in the lining application, or they were epoxy that had been applied over metal surface irregularities that looked like blisters.

The character of the blisters varied significantly among the five samples with blisters. Sample #1 had many blisters of varying sizes, some as large as 1 inch in diameter. It was unclear on this sample if the blisters accompanied the deterioration or if the deterioration resulted in the blisters. Most of the blisters had broken, and red iron oxide products were diffusing out of the blisters. Examination of the metal surface under the blisters showed a visible deposit of red and yellowish corrosion products.

Sample #5 had a few small isolated, closed blisters, ¼ inch in size. Samples #103 and #106 had a few large blisters, up to 1.5 inches in size. Sample #12 was difficult to evaluate for blistering because the relationship between the corrosion tubercle and blistering was unclear. Sample #12 did not show blisters on any areas of the coating without tubercles, in spite of the severe deterioration.

The coating was lifted from the metal surface to examine the surface under the blisters. Based on visual examination and the back surface of the blister, there did not appear to be any significant loss of metal. However, visual examination may be deceptive because the presence of graphitized cast iron or ductile iron could mask the actual metal loss. The surface was probed with a knife and cleaned with a power mini-brush to remove any loose corrosion products and graphite. There was no measurable material loss under the blisters, compared to the surrounding nonblistered area.

Dry Film Thickness

DFTs of the epoxy linings were measured by caliper, or in the case of sample #1, by stereomicroscope. Coating chips that had been pried from the pipe surface were measured randomly. When ringing was present, a similar number of measurements were taken on chips that represented the maximum and minimum thickness of the rings. Separate measurements were made and reported for any pools of lining observed in the pipe samples.

The average, median, maximum, and minimum DFT of each of the epoxy linings is shown in Table 4.3 and in Figure 4.1. The specification governing the lining application (WRc, 2001) calls for a minimum thickness of 40 mils, uniformly applied.

Of the 17 field samples, 1 had an average lining thickness in the range of 20 mils (sample #1 Yorkshire), 2 had a somewhat low DFT of 32 mils (samples #3 and #9 Yorkshire), and 2 were just below 40 mils (samples #5 and #12 at 37 and 38 mils, respectively). The remaining coatings had DFTs ranging from 44 mils to 60 mils.

All of the linings except samples #105 and #107 had minimum DFTs below 40 mils, indicating that some portion of the sample had insufficient DFT. Conversely, the thickest portions of the epoxy linings ranged up to 130 mils. The thickest areas of sample #1 were 27 mils, considerably below the required 40 mils.

The standard deviation of the DFT from average shows the wide variation in thickness due to surface roughness and/or ringing. Pooled areas of coating had DFTs exceeding 200 mils in some cases.

Volcanoes and Holidays

Many of the samples had visible volcanoes and holidays. Volcanoes consist of rounded cones of epoxy with a channel running through the middle of the cone to the metal substrate. Holidays are simply holes in the lining. Examination of the morphology of the volcanoes and holidays indicated that they formed during application of the lining. Volcanoes and holidays form when the metal surface contains a solid or contaminant that repels the coating. Holidays may also form from large foam cells, which remain trapped within the lining as it begins to gel.

The number of holidays was determined quantitatively on the split sample for each epoxy lining (ASTM D5162). A high voltage holiday detector enabled both visible and invisible holidays to be located and circled for counting and further study. The number of holidays per square foot was calculated, and the results are shown in Figure 4.2. The size of the holidays ranged from barely visible to about 1 mm in size, including volcanoes.

Table 4.3
DFT of historical field samples

	Pipe Sample Number																
Parameter	1	3	4	5	6	8	9	10	11	12	101	102	103	104	105	106	107
Maximum DFT, mils	27	44	72	47	88	60	44	134	101	49	88	132	87	61	68	62	69
Minimum DFT, mils	6	27	34	29	25	30	18	20	35	31	33	35	19	38	44	27	40
Standard deviation	8	4	11	5	19	11	9	41	16	5	16	27	20	5	8	9	8
Average, mils	19	32	47	37	47	44	32	62	56	38	54	63	53	47	52	49	49
Median, mils	21	32	48	37	45	44	33	50	56	38	51	55	51	47	49	49	47
Ringing	Yes	No	No	No	No	Yes	No	Severe	No	No	No	Large	Yes	Yes	No	No	Strong
Pools	No	216	No	No	No	203	103	No	No	No	No	No	No	No	No	No	146

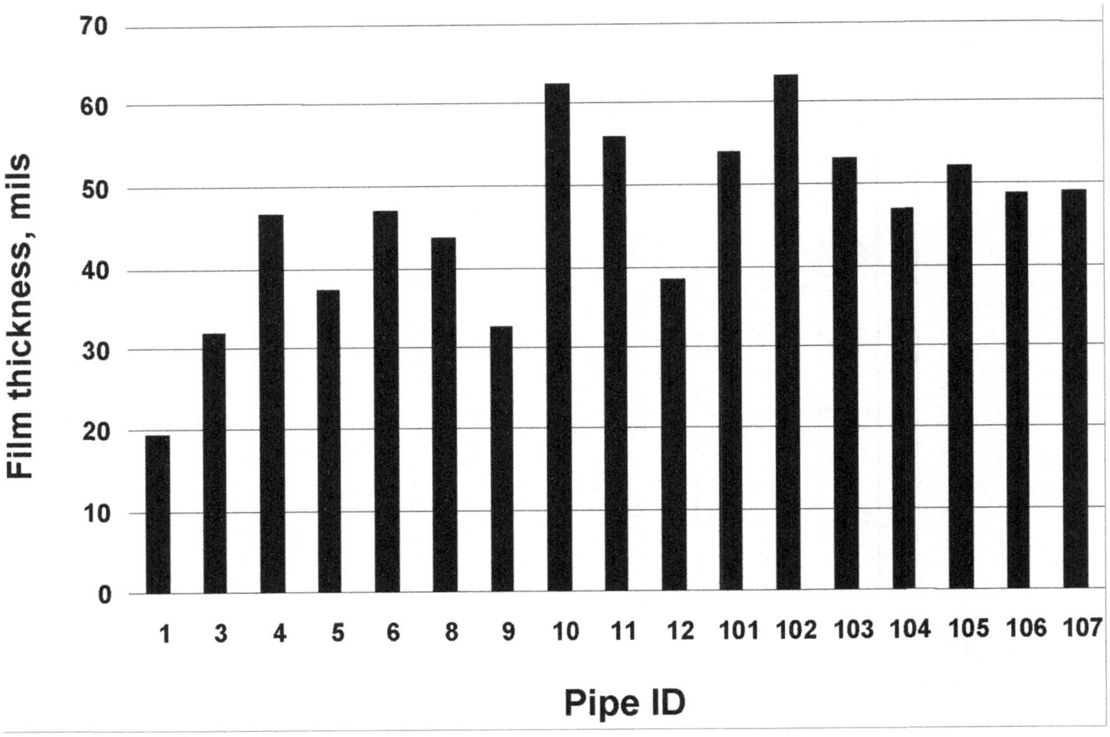

Figure 4.1 Average DFT of various samples

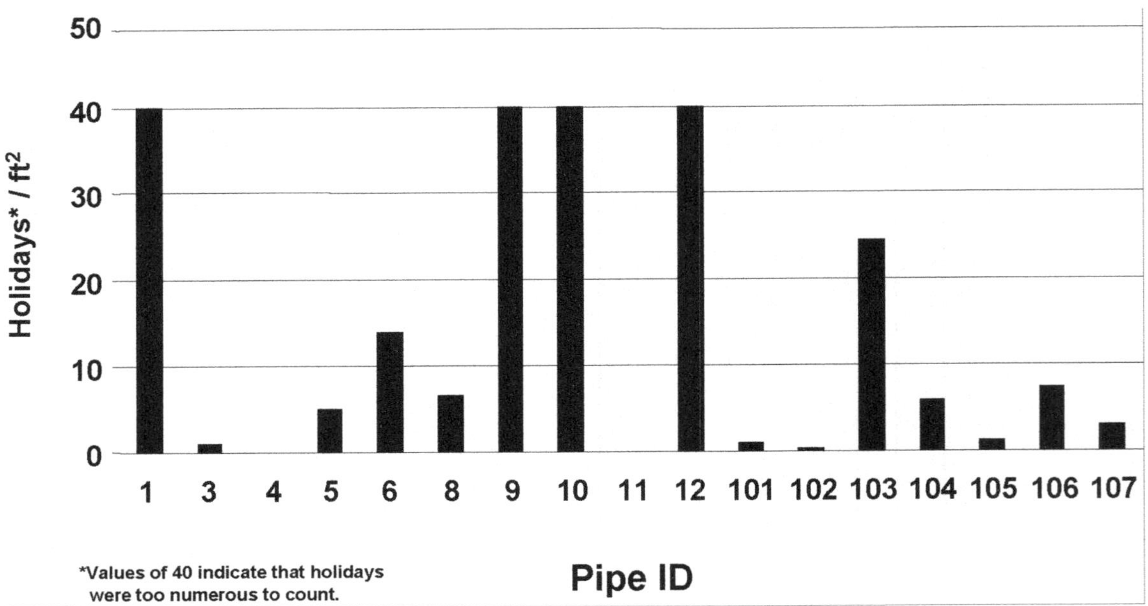

Figure 4.2 Holiday densities of various samples

Only 2 of the 17 samples, #4 and #11, were holiday-free. Six of the 17 samples had 5 or fewer holidays per square foot (samples #3, #5, #101, #102, #105, and #107). In three samples, the holiday count was described as continuous because there was either no holiday-free area on the lining (samples #1 and #12), or the exact number of holidays was too large too count (sample #106). Sample #10 had a large area (1 inch by 10 inches) with continuous holidays.

Sample #5 had a tightly adhering layer of conductive corrosion products over about half of the pipe surface. Visual inspection suggested that the underlying lining was holiday-free, or mostly holiday-free. Sample #105 had a similar longitudinal strip of corrosion product deposition.

Some samples had longitudinal lines of holidays (e.g., sample #8) due to an underlying irregularity on the metal surface (a crevice with a ridge). The result was that the epoxy was unable to bridge the irregularity, resulting in a gap or string of holidays.

Relationship of Holiday Density with Dry Film Thickness

Holiday density values measured for various samples were plotted with DFT and are shown in Figure 4.3. Although the data in Figure 4.3 are scattered, there is a trend of data that shows that larger DFT may result in less holidays.

Knife Adhesion by Parallel Scribe

The adhesion of the epoxy linings was evaluated by knife adhesion as described in American Society for Testing and Materials (ASTM) D6677. A parallel scribe cut was used rather than the X-cut specified in the ASTM standard because it provided more area for adhesion assessment, and an easier assessment of chip size. Parallel cuts, $\frac{3}{16}$ inches apart, were made through the lining to the base metal with a Dremel tool cutting disk. The lining between the scribe marks was pried with a utility knife and the adhesion was rated on a scale from 1 to 10 given in the standard. The scale is based on the difficulty of removing the coating and the size of the chips, and is shown in Table 4.4.

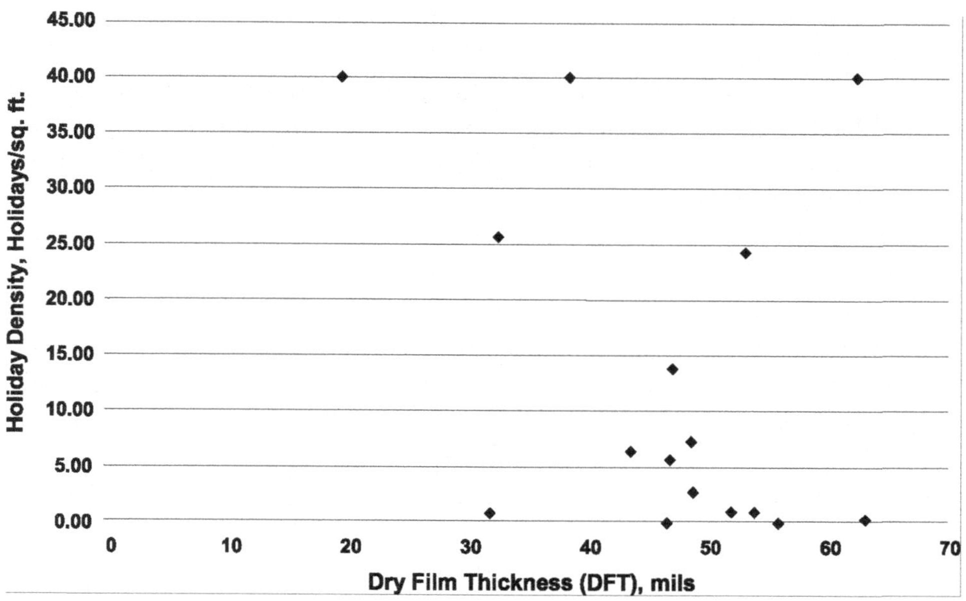

Figure 4.3 Holiday density versus DFT

Table 4.4
Scale for measurement of adhesion

Rating	Description
10	Coating is extremely difficult to remove; fragments no larger than approximately 0.8 by 0.8 mm ($\frac{1}{32}$ by $\frac{1}{32}$ inch) removed with great difficulty.
8	Coating is difficult to remove; chips ranging from 1.6 x 1.6 mm ($\frac{1}{16}$ by $\frac{1}{16}$ inch) can be removed with difficulty.
6	Coating is somewhat difficult to remove; chips ranging from 3.2 x 3.2 mm ($\frac{1}{8}$ by $\frac{1}{8}$ inch) to 6.3 x 6.3 mm ($\frac{1}{4}$ by $\frac{1}{4}$ inch) can be removed with slight difficulty.
4	Coating is somewhat difficult to remove; chips in excess of 3.2 x 3.2 mm ($\frac{1}{4}$ by $\frac{1}{4}$ inch) can be removed by exerting light pressure with a knife blade.
2	Coating is easily removed; once started with the knife blade, the coating can be grasped with the fingers and easily peeled to a length of at least 6.3 mm ($\frac{1}{4}$ inch).
0	Coating can be easily peeled from the substrate to a length greater than 6.3 mm ($\frac{1}{4}$ inch).

The adhesion of the lining in the split samples was evaluated by knife adhesion, using the parallel scribe method used for the laboratory panels in the immersion test. The adhesion ratings of the linings ranged from 0 to 4, with most ratings in the range of 0 to 2. The linings with the highest adhesion were samples #3, #4, #6, and #9, with ratings ranging up to 3 and 4. On each sample, the adhesion ranged from place to place, and with DFT due to the effect of these factors on prying strength. Figure 4.4 shows the knife adhesion rating of various linings.

Under Film Corrosion

The epoxy lining was pried away from the pipe surface so that the pipe surface could be examined. The surface under the lining consisted of a layer of black material (presumably graphitized steel and/or black iron oxide corrosion products) over a hard metal substrate. The thickness of the black material varied from sample to sample, and from area to area on one sample. Examination of the cut edges of the pipe showed that the graphitization penetrated 10 to 15% of the wall thickness in some areas of some samples. It was not possible to determine how much of the graphitization had been present at the time of lining application. The black material was soft and could be scraped from the surface with a utility knife. Red corrosion products were seen in minor amounts on only a few samples (#5, #6, #10, and #12), and were normally associated with blistered areas. Upon standing exposed to the laboratory environment, a few of the black surfaces developed a reddish look, presumably from oxidation of iron oxide products to red ferric oxide.

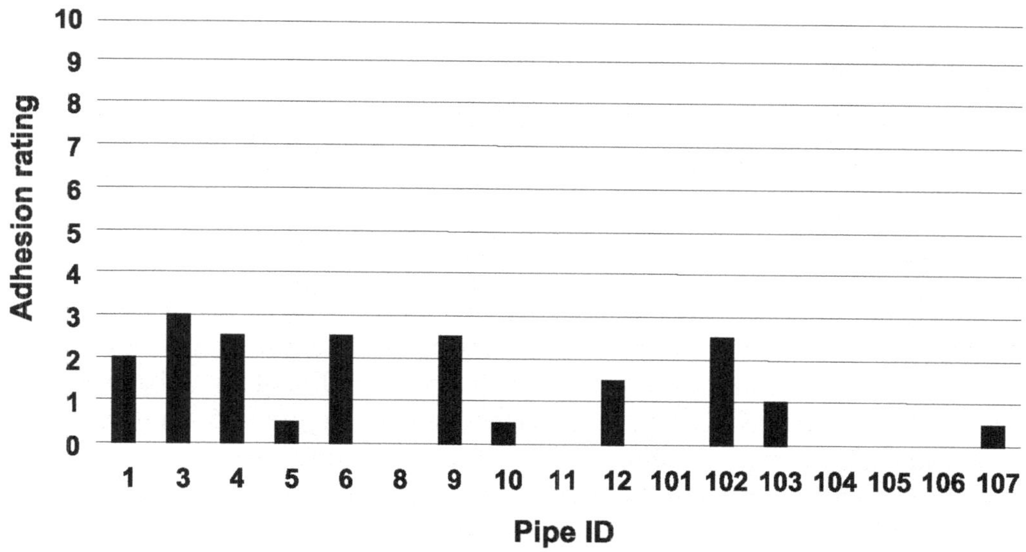

Figure 4.4 Knife adhesion rating for various linings

Various markings presumably from scraping and cleaning equipment could be observed on the pipe surface under the various coatings. The markings consisted of shallow scrapes and striations, the morphology of which was reflected on the back of the epoxy chips pried from the surface. In some cases, the structure of the original cast surface was present, with minimal graphitization. The ability of the epoxy lining to produce a replica of the underlying metal surface indicates that it generally flows out well and makes excellent physical contact with the surface, a highly desirable attribute. In most cases there was no discernible metal loss. In several instances metal loss was visible, but it was too small to measure on the rough surface.

Corrosion at Lining Defects

Most of the epoxy linings showed either very little or no rusting associated with holidays or blisters, with the exception of samples #1, #12, and #106. Most of the holidays in the linings, including the large holidays, were not accompanied by any rust stains around the holidays from the underlying metal. However, many of the epoxy linings had rust stains, streaks, and speckles on their surface, presumably from iron oxides originating upstream from unlined piping.

Square cuts were made through the lining surrounding a pinhole on various samples. This enabled the piece of coating to be pried from the surface in one piece, so that the metal area immediately below the pinhole could be examined for accelerated loss of metal. There were no red corrosion products associated with the pinhole, and there was no visible loss of material for the various pinholes examined. However, the graphitized residue may mask the corrosion that is actually occurring on the intact cast iron underlying the graphitized layer. Therefore, on selected holidays the graphitized layer was scraped away so that the underlying cast iron could be examined for localized metal loss under the holiday. In most cases there was no metal loss. In one case, there was a small amount of metal loss observed, but it was too small to measure.

Asphalt Lining Residues

One sample (#6) had a significant amount of asphalt residue, which did not appear to interfere with the function of the epoxy lining. The adhesion to the asphalt was not noticeably different than to the cast iron.

EIS Measurements

The impedance of the epoxy linings was determined using the EIS technique (ISO 16773). EIS provides a measure of barrier properties of the lining to water-based process fluid and corrosion current. Deterioration of the effectiveness of lining as a barrier is measured accurately from the reduction of EIS impedance values. Figure 4.5 shows a measure of interpretation of impedance in terms of barrier properties and protectiveness.

The methodology was the same as that used for the laboratory immersion test panels. A short section of acrylic pipe (1.25 inches in diameter) was cemented to the coating surface, and filled with tap water to hydrate the lining at room temperature. If the lining was disbonded, conductive gel was placed between the lining and the cast iron or an aluminum substrate to facilitate the measurement. Intact areas of the coating, with DFTs close to 40 mils where possible, were selected for EIS measurements. EIS measurements were made after intervals of 1, 2, and 3 weeks of hydration. The results are shown in Figure 4.6.

There was no significant difference in the impedances that were measured for the respective coatings at hydration intervals of 1, 2, and 3 weeks, indicating that hydration of the liners was essentially complete after 1 week (Walker and Guan, 1997).

The impedance of all of the liners was greater than log Z = 10.5, except for samples #1 and #12, both of which were in a highly deteriorated condition. Pipe sample #1 had an impedance of log Z = 5.4, and pipe sample #12 had an impedance of log Z = 7.3. Although it appeared to be highly deteriorated, sample #106 had an impedance of log Z = 10.8, indicating that the barrier properties of the lining were excellent where the lining was intact.

Cement Lining

Two of the samples (#101 and #107) contained a cement lining, over which the epoxy coating had been applied. Based on visual inspection, the presence of the cement lining did not have any obvious effect on the application quality or performance of the epoxy lining. Examination of the underside of the epoxy coating showed that it flowed out well over the cementation surface. The adhesion to the cementation surface was less compared to the cast iron, not unexpected for an alkaline surface. In the case of sample #101 the entire epoxy lining from the split sample could be removed in one piece. The holiday density and surface irregularities were not noticeably different from the other liners applied to cast iron. There were no blisters.

SUMMARY

Historical epoxy linings, in general, were found to be in excellent condition with excellent barrier properties except two samples. Adhesion has been found in general to be low. Various application defects, including holidays and variations in DFT, were found. Holidays were present in most linings but they did not result in significantly accelerated metal loss. Blisters were rare and similarly, were not associated with accelerated metal loss.

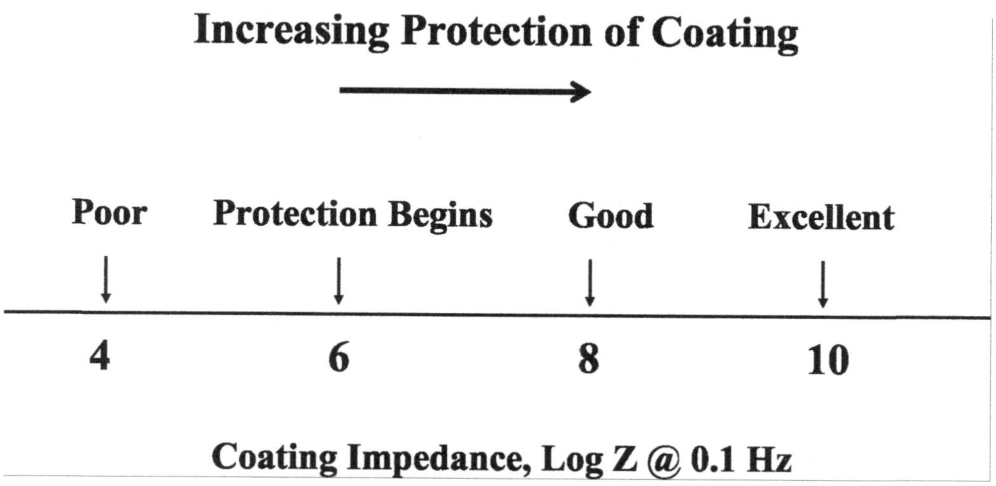

Figure 4.5 Interpretation of coating impedance as a barrier

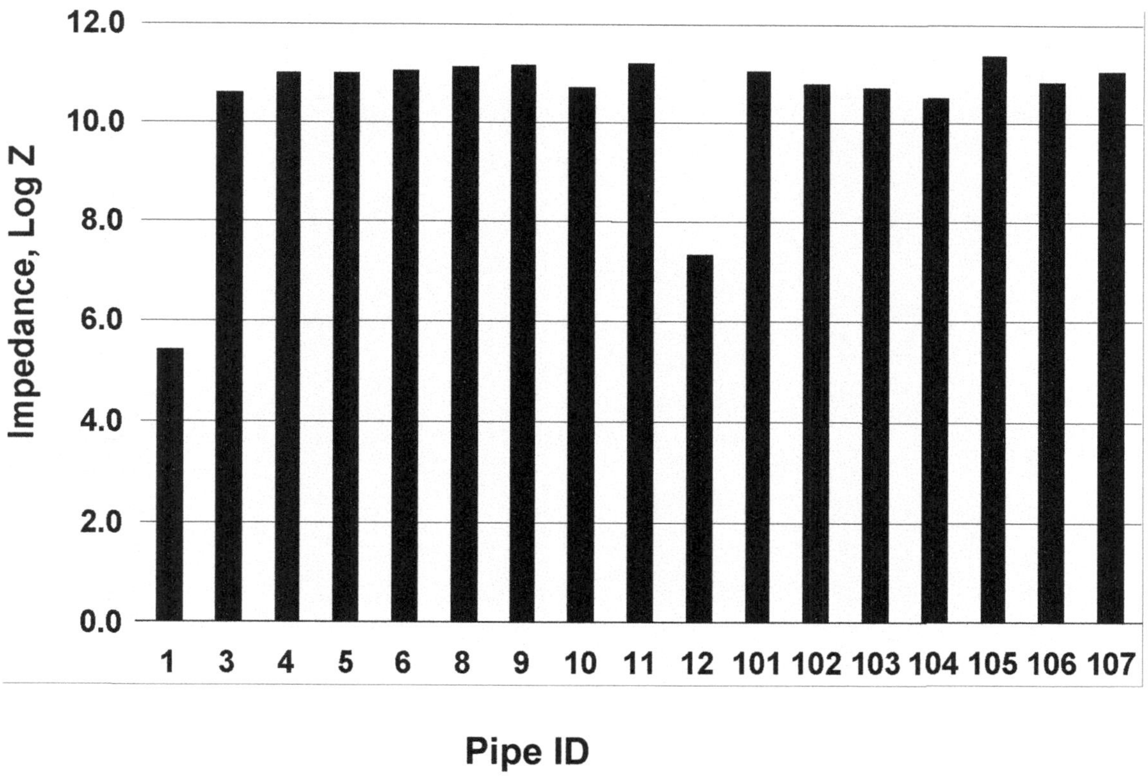

Figure 4.6 Average impedance (log Z) for various linings

CHAPTER 5
PARAMETERS AFFECTING PERFORMANCE OF EPOXY LINERS

INTRODUCTION

Data collected on various parameters from laboratory analysis of field pipe samples were analyzed to identify important parameters affecting the performance of epoxy liners. For each epoxy the following parameters were analyzed:

- DFT
- Holiday density
- Knife adhesion
- EIS measurements
- Physical observation of condition of the pipe
- Construction parameters (preparation of pipe such as cleaning and drying of pipe, application control, QA/QC of application)

Altogether 17 historical pipe samples with epoxy linings were collected from 5 water utilities in North America and the United Kingdom consisting of 8 different Type 1 and Type 2 epoxy materials.

ANALYSIS OF LABORATORY DATA

For all field pipe samples, DFT, holidays, knife adhesion, and impedance (EIS) of liners were measured, and physical conditions were observed and noted. All relevant information derived from all laboratory-measured data, including DFT, holiday density, knife adhesion, and EIS, are given in Chapter 4.

EIS is a parameter that measures the effectiveness of the liner in serving as a barrier for water to come in contact with the metal surface of a water pipe. High EIS data are indicative of a good barrier between water and the pipe metal. Pipe sample #1 and pipe sample #12 have low EIS values, indicating poor barrier condition of the liners. EIS values have been measured in locations on the liners that were free of holidays. As a result, EIS data may not show the condition of the liner as a whole.

Holidays in the liner play an important role in the life of a liner. Even if the epoxy material is good, but due to poor installation of the liner, may result in holidays, water then may come in contact with the pipe metal and reduce the effectiveness of the purpose of the liner. Holiday density depends on lining construction effectiveness, DFT, and surface preparation. For a well-constructed lining, holiday density should be close to zero.

Knife adhesion is a parameter that indicates the adhesion of the lining with the wall of the water pipe. Knife adhesion depends largely on the preparation of the surface and also on the epoxy material.

In order to develop criteria for the performance of liners, marginal values for each of these parameters beyond which liner performance may be questionable were identified and are as follows:

- Average lining thickness (DFT) less than 40 mils
- Holiday density more than 20 holidays per square foot
- Knife adhesion value of less than 1
- EIS value (log Z) less than 8 ohm-cm^2

For all of the pipe samples, Table 5.1 shows pipes with these deficiencies.

Table 5.1
Pipe sample marginal parameter values

Sample No.	Avg. DFT less than 40 mil	Min. DFT less than 20 mils	Holiday density more than 20 holidays/sq. ft.	Knife adhesion less than 1	EIS less than 8 ohm-cm^2	Visual inspection rating	Count
1	x	x	x		x	Poor	4
3	x				x	Excellent	1
4						Excellent	0
5	x			x		Excellent	2
6						Excellent	0
8				x		Excellent	1
9	x	x	x			Excellent	3
10		x	x	x		Medium	3
11				x		Excellent	1
12	x		x		x	Poor	3
101				x		Excellent	1
102						Excellent	0
103		x	x			Fair	2
104				x		Fair	1
105				x		Good, although there is a heavy film of corrosion products over the surface	1
106				x		Poor	1
107				x		Good	1

Table 5.1 indicates pipes with holiday density more than 20 holidays per square foot. These pipes are samples #1, 9, 10, 12, and 103. Holidays for samples #1, 10, and 12 are found to be continuous. In addition, two other samples, samples #9 and 103, show high holiday density.

Table 5.1 shows that most of the pipes have knife adhesion values less than one. Pipe samples #1, 9, 10, and 103 have an average DFT less than 20 mils. Two pipe samples, #1 and #12, have EIS values less than 8 mils, indicating poor barrier properties.

Considering all of the parameters, pipe samples #1 and 12 failed in four and three categories, respectively, which indicates poor condition of these liners. Pipe samples #4, 6, and 102 are in excellent condition without failing in any category.

Conceptual Lining Performance Analysis

Figure 5.1 shows a conceptual diagram of a liner adherence with pipe. With no holidays in the lining, the only path through which water can reach the pipe metal is by penetrating the lining barrier. With good construction, with no holidays, high barrier property measured by EIS values is a very important parameter. Therefore, in an epoxy lining project, EIS measurement is the most important measure of the long-term service life of a lining. Therefore, with good construction and with high barrier property, a liner can last many years.

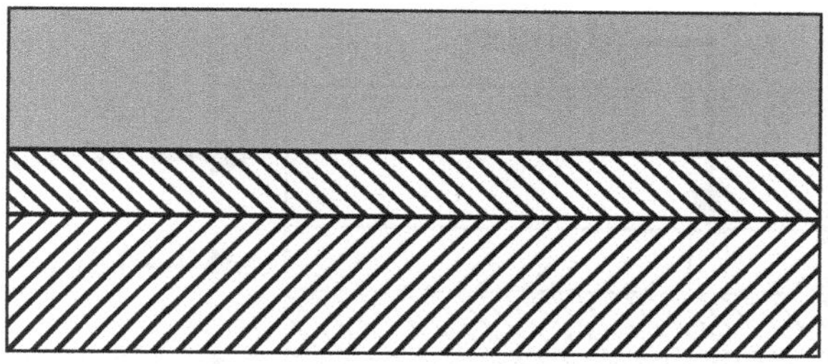

Notes: 1. Well-adhering liner with no holidays

2. Pipe is protected from water by barrier properties of liner (measured by EIS)

Figure 5.1 Conceptual diagram of liner adherence with pipe

Holidays in the liner play an important role in the life of a liner. Even if the epoxy material is good, but due to poor installation of the liner may develop holidays, water then may come in contact with the pipe metal and reduce the effectiveness of the liner.

Figure 5.2 shows a conceptual diagram of a liner with holidays. With holidays, water can pass through the holidays and come in contact with the pipe metal. Therefore, application procedures are very important to avoid holidays. However, if the adhesion of the epoxy liner with the pipe wall is good, water cannot creep under the lining and major rusting can be controlled.

Figure 5.3 shows the same conceptual diagram as shown in Figure 5.2, except it has poor adhesion. As a result of poor adhesion and holidays, water can pass through the holidays and also creep under the liner, which may result in extensive rusting of the pipe metal. Therefore, it appears that if holidays are present, adhesion is a very important parameter.

SUMMARY

From this analysis, it appears that the following parameters are important for determining the longevity of a lining in a water pipe:

- EIS measurements of the lining material
- DFT
- Holiday density
- Knife adhesion

EIS depends on the property of the epoxy. Most of the recent epoxies have high EIS values. DFT, holiday density, and knife adhesion depend largely on QC of construction. Therefore, it is of utmost importance that the construction of a liner is performed with proper QA/QC and strictly follows the guidelines and specifications.

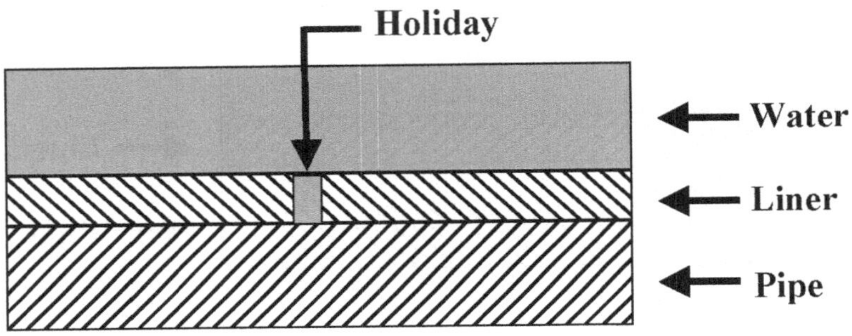

Notes: 1. Well-adhering liner with no holidays

2. Water contacts pipe at holidays, but contact area may not progress due to good adhesion

Figure 5.2 Well-adhering liner with holidays

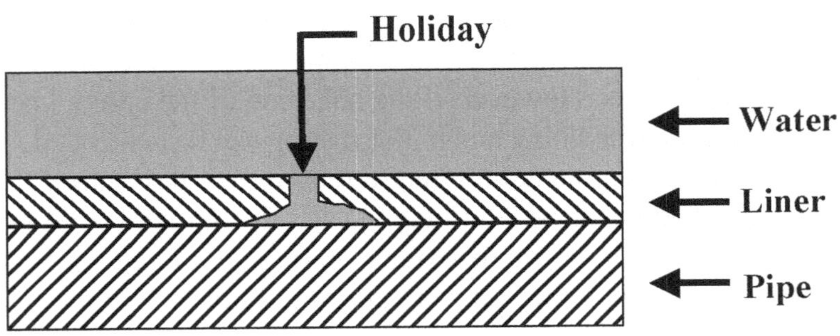

Notes: 1. Poorly adhering liner with a holiday

2. Water contacts pipe at holidays, and poor adhesion allows water and corrosion to spread under liner (undercreep)

Figure 5.3 Poorly adhering liner with holidays

CHAPTER 6
LONG-TERM ACCELERATED TESTS

INTRODUCTION

Epoxy linings are routinely applied to steel structures to control corrosion. The epoxy lining serves as a barrier between the steel surface and the service environment, protecting the steel from corrosive process fluids, water, corrosive gas, etc. The epoxy lining slowly degrades with time, until it is no longer able to protect the steel. The rate of deterioration depends on the nature of the service, the type of epoxy, and the quality with which the lining was applied. Surface preparation in particular has a strong effect on lining longevity, and a great deal of emphasis is placed on thorough cleaning and profiling of the surface prior to lining application.

Water-line rehabilitation with epoxy and polyurethane linings is quite different in that surface preparation is by necessity limited, typically consisting of mechanical cleaning and swabbing. Thus, the surface may be moist and contain varying degrees of roughness and foreign material. Therefore, a different approach is needed to be taken for predicting lining longevity.

DEVELOPMENT OF TESTING METHODOLOGY

The ability to predict the service life of a protective lining is much sought after by lining manufacturers, facility owners, and researchers. Developing such predictive models has been a challenge, with only several notable successes, i.e., prediction of service life of coatings applied to steel for protection against outdoor weather, and the prediction of service life for coatings in water immersion by the United States Navy.

The approach is to expose test panels of the linings to a simulated service environment under controlled laboratory conditions. Obtaining suitable data relies on the following:

- Lined panels are exposed to a sufficiently aggressive version of the service environment, such that the rates of deterioration are suitably accelerated.
- The simulated service environment includes all of the relevant parameters controlling deterioration.
- The deterioration processes and mechanism, although more rapid, are the same as those in service.
- Lining performance (and deterioration) is assessed using suitable measurements, which characterize key attributes of deterioration.
- The rate of deterioration (based on physical or chemical characterization of the test panels) can be modeled and extrapolated.
- The economic end of the lining life can be defined, typically in quantitative terms of loss of protection to the steel substrate.
- Longevity can be derived based on "end points" for the performance parameters, which indicate the end of the lining life.

The accelerated test results are validated and/or adjusted based on field observations and experience. Linings that perform well in the field should perform well in the accelerated test, and linings that do not perform well in the field should reflect that in the test. Failure mechanisms in the field and in the laboratory should be similar.

The design of the accelerated test consisted of immersing test panels of the various lining preparations in water at three different temperatures for various intervals of time. The test was accelerated by using elevated temperatures, low-ionic strength water, reduced lining DFT, and intentional holidays. Because the rate and mechanism of deterioration was uncertain, a number of parameters were monitored for the purpose of evaluating lining performance during immersion. The data were later evaluated to identify those parameters most suitable for estimation of lining life expectancy.

The following parameters were followed as a function of immersion time and temperature to evaluate lining performance:

- DFT change
- Barrier properties, as evaluated by measuring EIS
- Blistering on a well prepared and a poorly prepared steel surface
- Hardness
- Knife adhesion on a well prepared and a poorly prepared steel surface
- Pull-off adhesion on a well prepared steel surface
- Undercreep at intentional mechanical defects on a well prepared and a poorly prepared steel surface
- Underfilm rusting and metal discoloration on a well prepared and a poorly prepared steel surface
- Wear resistance

Due to the limited surface preparation, the water line epoxy linings have limited adhesion. Therefore, a key part of their function was anticipated to be barrier properties, or the degree to which they prevent water from reaching the underlying metal. Barrier properties (or permeability to water) are governed by factors such as epoxy formulation, cure processes, and DFT. As linings age, their physical and chemical breakdown results in a gradual decrease in barrier properties until the lining is no longer able to protect the metal surface.

One way of evaluating barrier properties is by EIS, a technique in which the impedance of the lining to alternating current (ac) electricity is measured. EIS has been used extensively for evaluating lining performance, barrier properties, and deterioration with time (Gray and Appleman, 2003). The higher the impedance of the lining, expressed as log Z where Z is impedance, the better are its barrier properties and the lower its permeability to water.

Anticipated performance of a lining based on log Z is shown in Figure 6.1, which is derived from a large amount of laboratory and field work (Gray and Appleman, 2003).

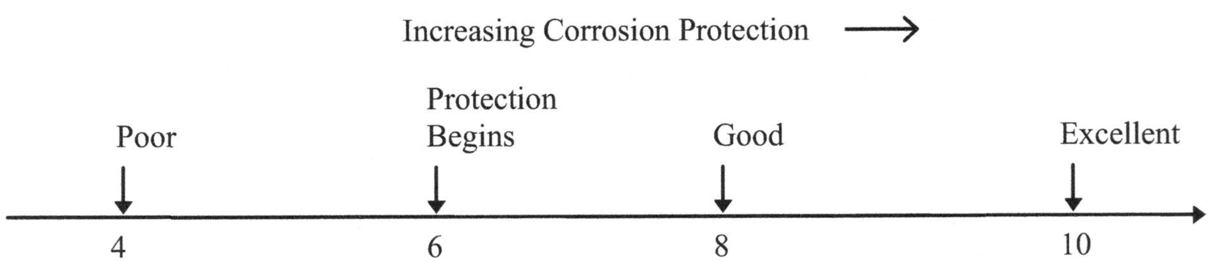

Figure 6.1 Corrosion protection of organic linings at various lining impedances at 25 °C, log Z (Z in ohms cm^2 at 0.1 Hz)

PREPARATION OF TEST PANELS

Four epoxy linings and one polyurethane lining were run in the test program. The epoxies are Nitoline WP, Copon 162PX, Geopox GX014, and ELC 257/91. They represented a range of epoxy brands, types, and eras of service. A rapid-set polyurethane lining, Copon Hycote 169, was also run in the test program.* Polyurethane has superseded the epoxies previously used for in situ relining in the United Kingdom. The four epoxies are coded A, B, C, and D.

The five linings were applied at a target DFT of 40 mils. Two of the epoxies (epoxy A and epoxy B) were also applied at a target DFT of 20 mils, to represent a compromised situation under which more rapid deterioration should occur. The actual average DFTs are shown in Table 6.1. Nitoline WP was obtained courtesy of EPCOR, an applicator in Vancouver, Brit. Col. Copon 162PWX was provided courtesy of the manufacturer, E. Wood, in the United Kingdom. ELC 257/91 was provided courtesy of Bucks County Water and Sewer Authority (BCWSA) in Pa. Geopox GX014 was provided courtesy of the manufacturer, SubTerra, in the United Kingdom.

Steel Test Panels

Steel test panels were cut from flat, hot-rolled plain carbon steel. Mill scale and minor amounts of corrosion scale were present on the surface. The test panels were 3 x 6 inches in size and 0.188-inch thick. The corners were radiused by grinding. A 2-mm hole was drilled $\frac{3}{16}$ inch from the top of each panel. Steel wire, run through the hole and tightly twisted, was used to mount the panels on bars for lining application. One corner was ground flat so that the two sides of each panel could be readily distinguished because the surface preparation was different for the two sides. A total of 80 panels were prepared for each combination of lining and DFT.

Steel tubes were also lined for the wear measurement portion of the program. These tubes were 0.78 inch in outside diameter and 16 inches long. Four tubes were prepared for each lining and DFT combination.

Table 6.1
Epoxy and polyurethane linings used in the test

Lining material	DFT (mils)	Type of epoxy
Epoxy A	20	1
Epoxy A	40	1
Epoxy B	20	2
Epoxy B	40	2
Epoxy C	40	1
Epoxy D	40	2
Polyurethane	40	

* The manufacturer of Copon Hycote 169 is E. Wood of the United Kingdom. E. Wood covered the costs for running the test, and the data are included in this report as a courtesy.

Surface Preparation of the Test Panels and Tubes

The surface preparation and lining application was conducted by a commercial applicator (CSI Coatings, Inc., Nisku, Alberta) for all of the linings, except polyurethane lining. The surface preparation and lining application for polyurethane was performed by the manufacturer.

Side A ("front" side) of each panel was solvent-cleaned, followed by blast-cleaning to a near white metal blast (Steel Structures Painting Council [SSPC] SP 10, National Association of Corrosion Engineers [NACE] 2) using a slag-based abrasive. The surface had a 3-mil (0.003 inch) profile based on analysis by Testex tape (ASTM D4417). This surface served as a highly uniform, high quality surface for which subtle differences in lining performance could be distinguished in adhesion, impedance, undercreep, and other parameters.

Side B ("back" side) of each panel was solvent-cleaned only (SSPC SP1) to remove any oil, grease, and soluble salts. Any loose corrosion products or mill scale were removed by steel brush prior to solvent cleaning. This side served as a surface that was more representative of field conditions in that it had an oxidized surface with residual scale and minimal surface profile.

The steel tubes were blast-cleaned to a near white finish.

After the panels and tubes were blast-cleaned, wired, and hung from the support rods, the panels were given a final blow-off with compressed air to remove any dust residues prior to lining application.

Application of Epoxy Lining

The panels were all lined by the same commercial applicator (CSI Coatings, Inc., in Nisku, Alberta) using the same equipment, which was a digitally controlled Graco plural component system (King pump). The component ratio could be set at the pump at ±0.1 unit. The Code of Practice was provided to the applicator, along with the applicable MSDSs, product data sheets, and application instructions from the manufacturers.

The two components for each of the respective linings were placed in 5-gallon containers fitted with band heaters for preheating. Preheat temperatures ranged from 35 °C to 40 °C, in compliance with the lining manufacturer's recommendations. The lining application was conducted indoors at temperatures of 18 to 25 °C. Cure temperature overnight ranged from 18 to 28 °C.

The panels were lined completely on all sides, including the wire hangers. The required DFT was obtained in a single application (one coat), except for epoxy C. The epoxy C started to run and sag after application of 20 mils; application was temporarily stopped for about 20 minutes to allow the lining to set up, after which a second application was made. This is within the allowable limit of 16 hours recommended by the manufacturer. The first batch of epoxy A liquid was preheated and the 40-mil DFT panels were prepared. Then a second batch of epoxy A was preheated and the 20-mil DFT panels were prepared. The 20-mil batch of panels showed some variation in color and had a somewhat rough finish compared to the 40-mil preparation. In the case of epoxy B, 20- and 40-mil DFT panels were sprayed at the same time using the same preheated material.

The four linings were applied over a period of 10 days as follows:

- September 9, 2004 (Thursday): Epoxy A 20 and 40 mils
- September 13, 2004 (Monday): Epoxy B 20 and 40 mils
- September 14, 2004 (Tuesday): Epoxy C 40 mils
- September 16, 2004 (Thursday): Epoxy D 40 mils

This schedule facilitated the preparation of the panels prior to being placed into immersion testing. It also facilitated the schedule for various post-immersion analyses that would be done over the course of the project.

The panels were blast-cleaned and mounted in the morning, followed by spray application of the lining between 2:00 p.m. and 3:30 p.m. in the afternoon. Film thickness was checked with a toothed wet-film thickness gauge. The panels were left to cure overnight. At 6:15 a.m. the next morning, the panels were removed from the racking, wrapped in foam wrap, and taken back to the laboratory where they were prepared for immersion testing. The wire hangers were snipped about ¼ inch above the top of the panel and left in place for testing. The lining fully sealed the hole and wire next to the panel. This enabled the panels to be placed into immersion 18 to 20 hours after spray application without delay due to repair or touch-up of the hanger area.

Photos of the panels, racking, and spraying are shown in Figures 6.2 and 6.3.

Observations After 16 Hours of Cure

After curing overnight (about 16 hours at 18 to 28 °C), the lining films were firm and could not easily be indented with a fingernail. However, most of the linings were quite sticky to the touch. Epoxy D and epoxy A were particularly sticky. Fingerprints and smudges were readily left on the surface during handling. Epoxy B was less sticky, while epoxy C had a smooth hard surface that showed stickiness only when pressed firmly with a finger.

EQUIPMENT AND PROCEDURES

Immersion Equipment and Procedures

The immersion test consisted of placing the test panels in tanks of distilled water. NACE TM0174 and ASTM 870 were used for guidance, with modifications to accommodate the needs of the current test program. Immersion tests were run at three temperatures: 8 °C, 25 °C, and 35 °C. The 25 °C and 35 °C temperatures were well above most buried line temperatures (5 to 15 °C), and therefore, served to accelerate any deterioration processes. Temperatures above 35 °C could not be used because the glass transition temperatures of the linings fall into this range. Distilled water was used rather than potable water to further accelerate the deterioration because the low ionic strength may promote further deterioration from adverse osmotic effects. Test panels were removed from the tank at intervals of 1 week, 1 month, 2 months, 4 months, and 8 months for evaluation of lining performance.

The immersion equipment consisted of covered glass tanks equipped with immersion heaters and pumps with aerators. This provided good mixing of water and air throughout each tank. The dimensions of the tanks were 12 x 11 x 29 inches. One tank was set up for each temperature. The 8 °C temperature was obtained by placing the water tank in a freezer equipped with a temperature controller. The variation in temperature within the tanks was 2 °C based on continuous monitoring and probing different parts of the tank with a thermocouple. Fresh distilled water was added to the tanks to maintain a constant volume as water evaporated or panels were removed.

Within the tanks, the panels were placed in polyvinyl chloride (PVC) racks, which gave the panels a slight tilt of 5 to 10 degrees. The panels were placed randomly within each tank, based on lining type and immersion time. Photos of the immersion tanks and panels are shown in Figure 6.4.

80 panels ready for coating

Blast-cleaned panels

Racking prepared steel panels

Solvent-cleaned panels

Figure 6.2 Panel preparation

Coated panels curing

Spray application of Geopox

Spray unit with heated reservoirs

Spraying ELC 257/91

Figure 6.3 Spray application of lining to panels

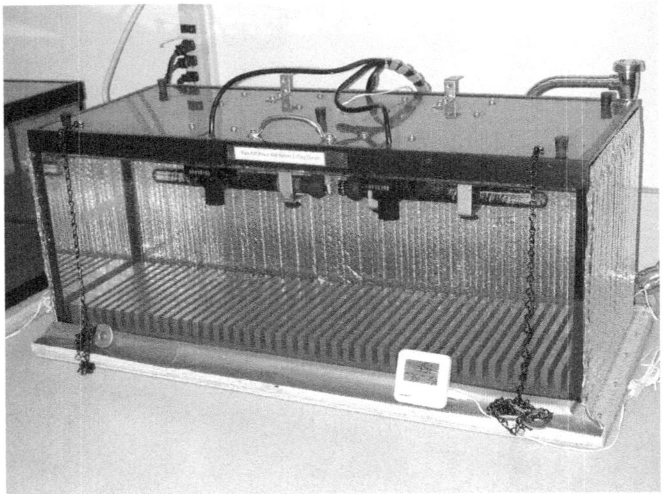

35 deg tank

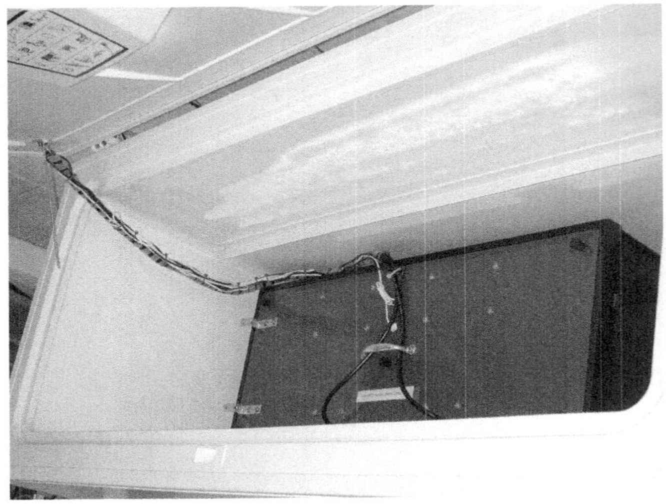

Freezer

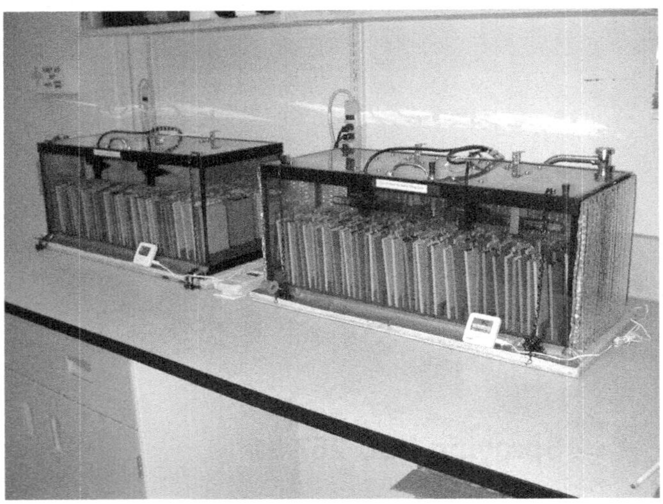

Full tanks 1

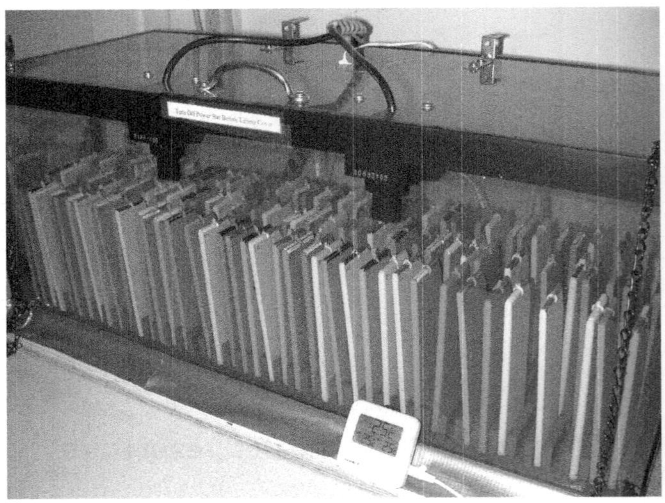

Full tanks 3

Figure 6.4 Testing apparatus and panels under test

Pre-Immersion Preparation of Lined Test Panels

Of the 80 panels that were prepared, the best were selected for testing in terms of target DFT and quality of application, i.e., a smooth uniform finish without runs, irregularities, and wet-gauge marks. The panels were sorted, labeled, and numbered. A total of 65 panels were put into the test, including immersion testing, impedance measurements, and untested controls.

The DFT was measured on both sides of each panel. This consisted of taking five individual DFT measurements, and recording the average DFT, standard deviation, maximum, and minimum values. The DFTs, shown in Table 6.1, were somewhat on the low side. For simplicity, the various applications will simply be referred to as the "20-mil" and "40-mil" applications.

Four panels of each lining (and respective DFTs where applicable) were specified for each of the respective combinations of immersion temperature and time. These panels were prepared as follows prior to immersion:

- Two panels (duplicates) were left fully intact, with sides A and B specified
- Two panels (duplicates) received a $\frac{1}{16}$-inch diameter holiday on both sides A and B, at the geometric center of the panel.

Side A was blast-cleaned, and side B was solvent-cleaned only. The purpose of the holidays was to assess adhesion and corrosion next to a defect, and determine the extent to which water can penetrate under the lining and disbond the lining.

Post-Immersion Evaluation of Lining-Panel Assignment

Upon completion of the specified immersion, the four panels for each lining were removed and the following tests were conducted, distributed as follows among the panels:

The two intact panels were evaluated for:

- DFT change (side A)
- Parallel scribe knife adhesion at the top of each panel (sides A and B)
- Discoloration and under-film corrosion as applicable (sides A and B)
- Buchholz hardness (side A)
- Pull-off adhesion (side A)
- Blistering (sides A and B)

The two panels with holidays on both sides were evaluated for:

- Discoloration (sides A and B)
- Undercreep (sides A and B)
- Under-film corrosion (sides A and B)
- Blistering (sides A and B)

Performance Evaluation of the Linings in Immersion

Various procedures were used to collect data about the performance of the linings as a function of immersion time and immersion temperature. Performance parameters and the methodology used in the measurement are shown in Table 6.2. Details of the measurement procedures are given in Appendix D. Details of performance tests and panel types are shown in Table 6.3.

Table 6.2
Parameters measured and methodologies used

Parameter	Methodology
DFT	ASTM D1186
Knife adhesion	ASTM D6677
Undercreep around holiday	ASTM D1654 (modified)
Corrosion of the steel surface	ASTM D610
Blistering	ASTM D714
Pull-off adhesion	ASTM D4541
Buchholz hardness	ISO 2815
Wear test	ASTM G6

Table 6.3
Grid showing performance tests conducted on panel types and sides

	Intact panel		Panel with holidays	
Measure of performance	Blast-cleaned side A	Solvent-cleaned side B	Blast-Cleaned side A	Solvent-cleaned side B
DFT	X		X	
Knife adhesion – parallel scribe	X	X		
Blistering	X	X	X	X
Undercreep			X	X
Corrosion of steel under coating	X	X		
Corrosion of steel around holiday			X	X
Hardness	X			
Pull-off adhesion	X			

TEST RESULTS

The results for each test are summarized here, along with various summary graphs and charts.

Dry Film Thickness

Measured DFT values are shown in Table 6.4. There was no significant change in the DFT of any of the linings, and therefore, no significant swelling, erosion, or shrinkage of the lining films.

Table 6.4
DFT (mils) 1 mil = 0.001 inches

KTA panel no.	Coating name	Generic type	DFT			
			Avg.	Std. dev.	Max.	Min.
AJZ100	Epoxy A 20 mils	Type 1	22	2	24	19
AJZ200	Epoxy A 40 mils	Type 1	39	2	42	36
AJZ300	Epoxy B 20 mils	Type 2	20	1	21	18
AJZ400	Epoxy B 40 mils	Type 2	36	2	38	34
AJZ500	Epoxy C 40 mils	Type 1	42	2	44	40
AJZ600	Epoxy D 40 mils	Type 2	39	2	41	36
AJZ700	Polyurethane	Polyurea	48	4.2	54	43

Knife Adhesion

The results are summarized in Figures 6.5 through 6.18. Results indicate that knife adhesion is higher on blasted surfaces than on non-blasted surfaces. Epoxy D and polyurethane show very low adhesion after 8 months, even on blasted surfaces.

Blast-Cleaned Side (Side A)

Epoxy C retained excellent adhesion on the blast-cleaned side of the panel, and had significantly better adhesion than any of the other linings. Immersion temperature and time had a nominal effect. The adhesion values after 8 months at the 3 immersion temperatures ranged from 8 ($\frac{1}{16}$-inch chips can be removed with difficulty) to 9.5 (fragments $\frac{1}{32}$ inch to $\frac{1}{16}$ inch removed with great difficulty).

The adhesion of epoxy B after 8 months of immersion at 8 °C and 25 °C was rated at 7 or higher, with the exception of the 20-mil film at 25 °C, which was rated 5. At 35 °C, the adhesion was somewhat lower but still relatively good, with ratings of 7 (20-mil film) and 5 (40-mil film) after 8 months of immersion.

The 40-mil film of epoxy A appeared to have less adhesion than the 20-mil film. Polyurethane film by nature has lower adhesion compared to epoxy linings.

Within 2 months of immersion, the adhesion of epoxy D had dropped to 1 (35 °C), 4 (25 °C), and 5 (8 °C). After 4 months the adhesion at all temperatures had dropped to 3 or less.

Non-Blast-Cleaned Side (Side B)

The adhesion of the linings was generally poor, as expected due to the poor surface preparation. The adhesion ratings of epoxy C after 8 months of immersion were relatively low, in the 2 to 3 range. Epoxy D and polyurethane had the lowest adhesion, based on the adhesion rapidly dropping to 0 to 1.5 within the first few months of immersion. The remaining linings had intermediate behavior. In general, adhesion decreased more rapidly as the immersion temperature was increased.

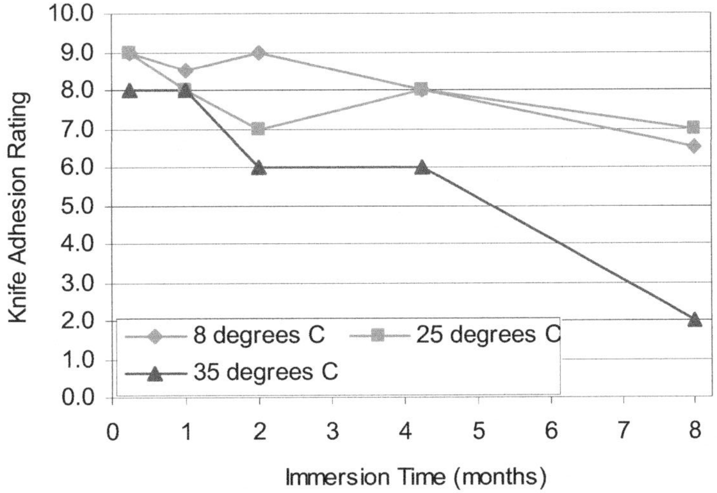

Figure 6.5 Knife adhesion results - blasted side (A) - epoxy A 20 mils

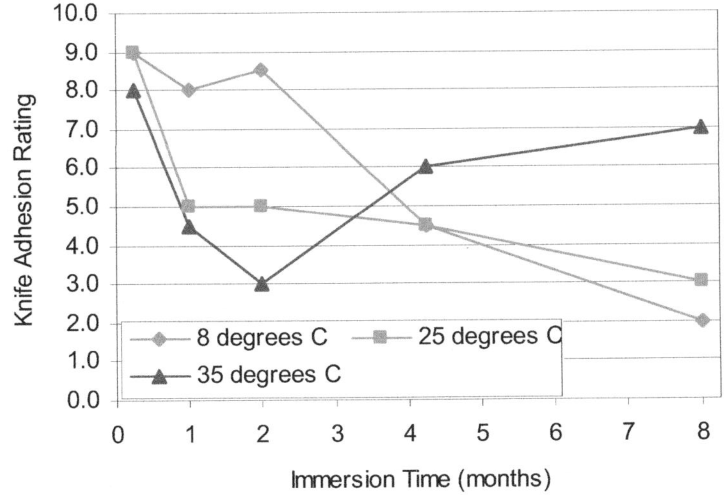

Figure 6.6 Knife adhesion results - blasted side (A) - epoxy A 40 mils

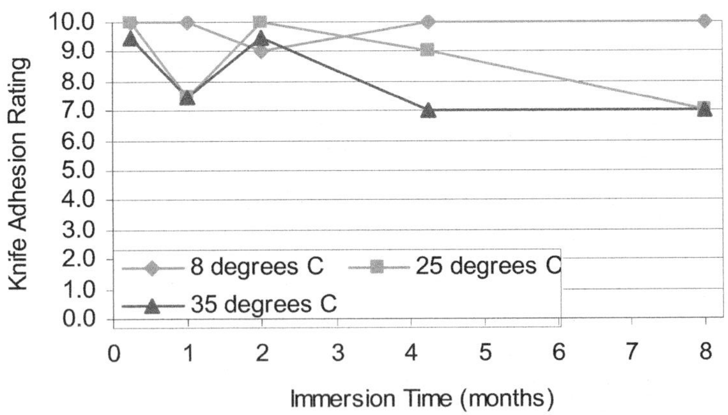

Figure 6.7 Knife adhesion results - blasted side (A) - epoxy B 20 mils

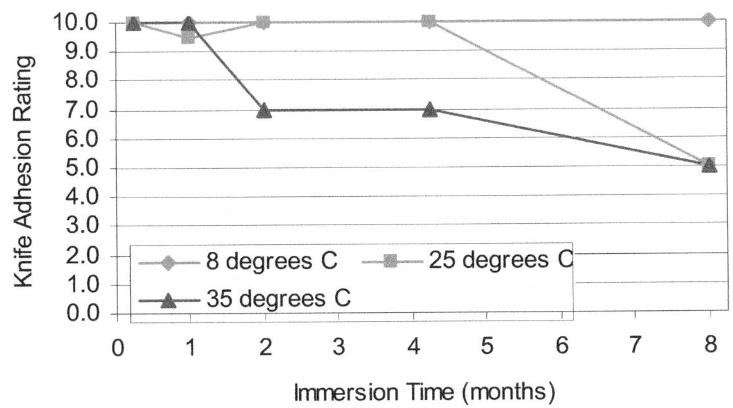

Figure 6.8 Knife adhesion results - blasted side (A) - epoxy B 40 mils

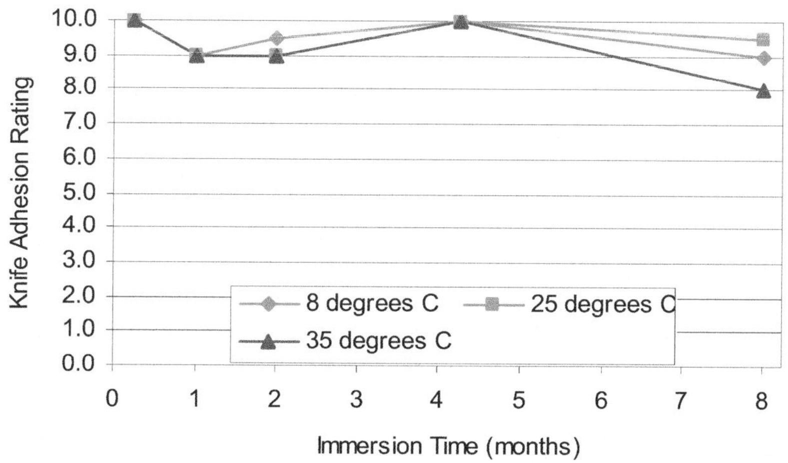

Figure 6.9 Knife adhesion results - blasted side (A) - epoxy C 40 mils

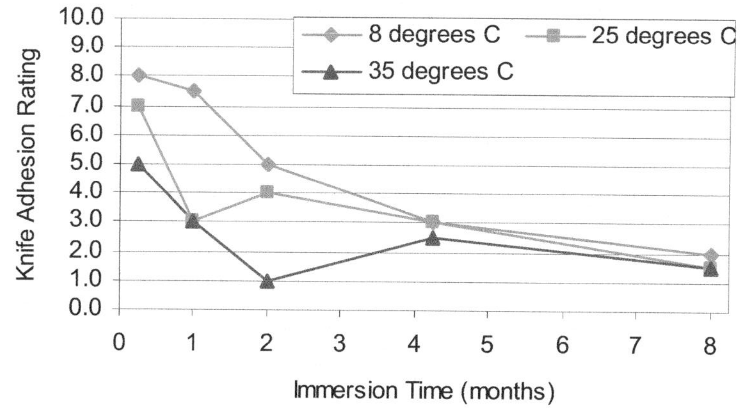

Figure 6.10 Knife adhesion results - blasted side (A) - epoxy D 40 mils

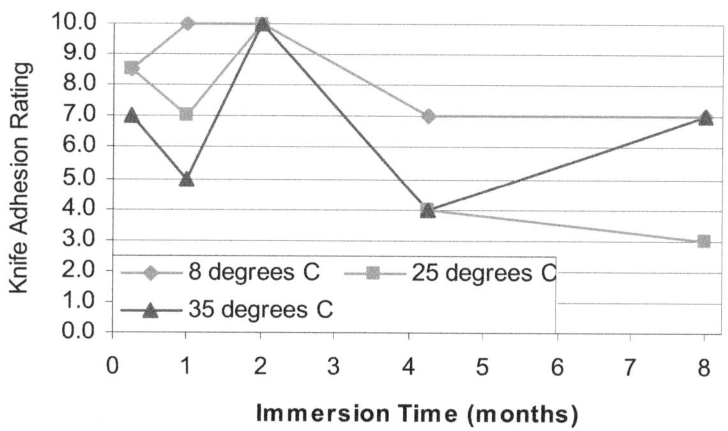

Figure 6.11 Knife adhesion results - blasted side (A) - polyurethane 40 mils

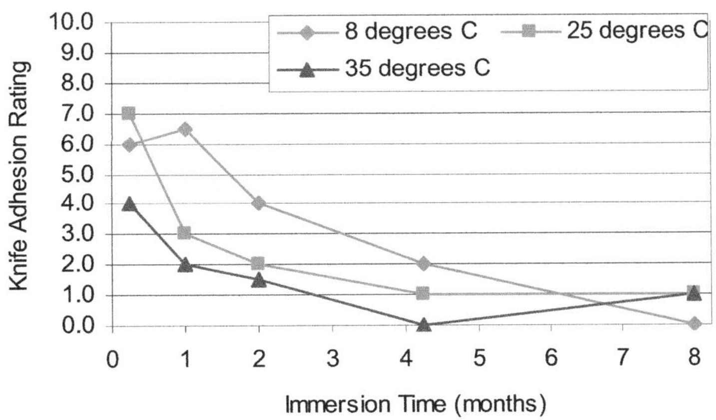

Figure 6.12 Knife adhesion results – non-blasted side (B) - epoxy A 20 mils

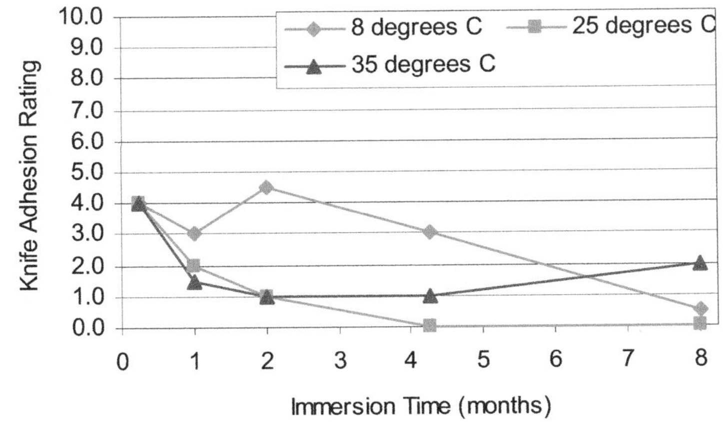

Figure 6.13 Knife adhesion results – non-blasted side (B) - epoxy A 40 mils

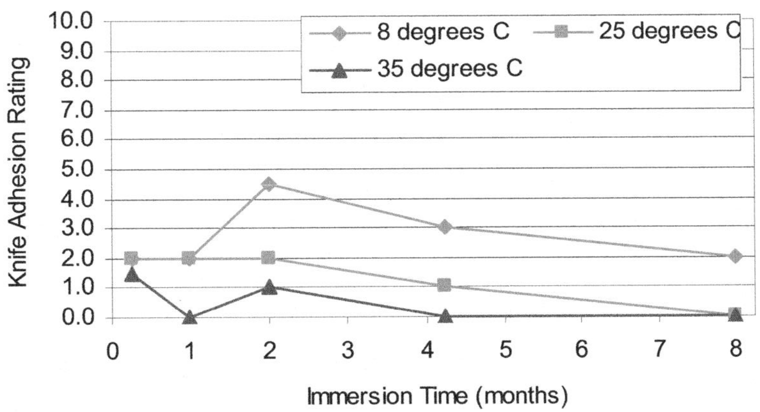

Figure 6.14 Knife adhesion results – non-blasted side (B) - epoxy B 20 mils

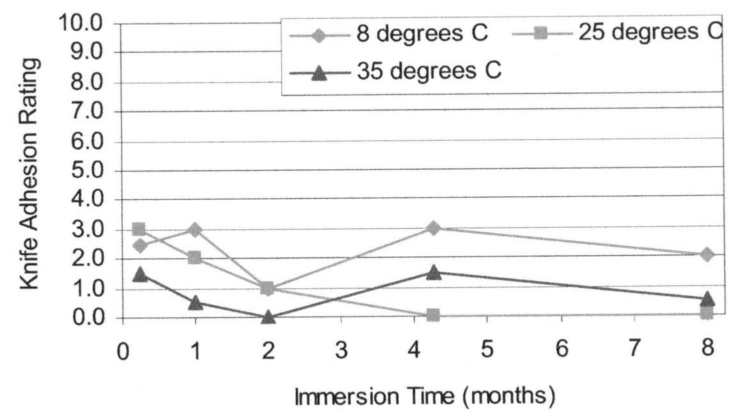

Figure 6.15 Knife adhesion results – non-blasted side (B) - epoxy B 40 mils

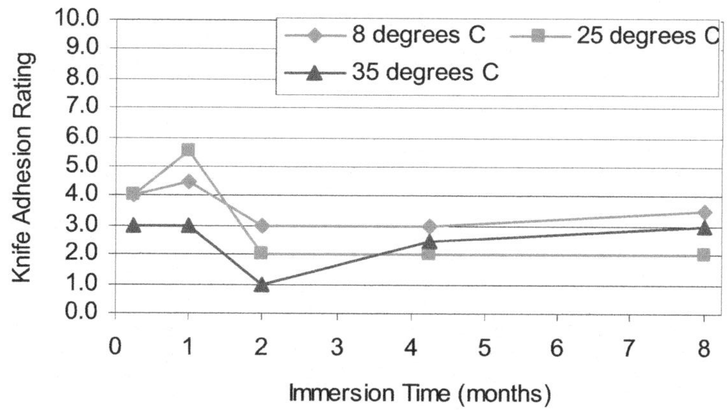

Figure 6.16 Knife adhesion results – non-blasted side (B) - epoxy C 40 mils

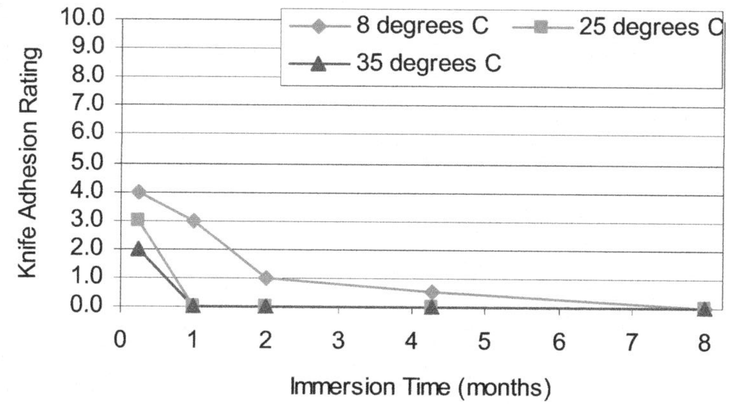

Figure 6.17 Knife adhesion results – non-blasted side (B) - epoxy C 40 mils

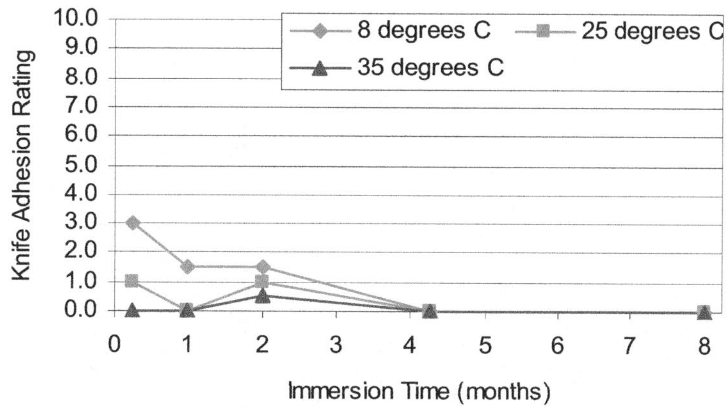

Figure 6.18 Knife adhesion results – non-blasted side (B) - polyurethane 20 mils

Discussion of Adhesion Results

Epoxy A remained relatively soft (and sticky) over the test duration. Typically, a utility knife cut through the lining, and only a limited prying action could be applied. In comparison, a strong prying action could be applied to the epoxies and polyurethane due to their high mechanical strength. The strong prying action enabled them to be more readily lifted from the metal surface, giving them a deceptively lower adhesion rating.

In general, the adhesion decreased with time and with increasing immersion temperature, but the trends were not consistent for a given lining or between linings. This is partly due to the inherent subjectivity in the test, related to prying action and the rating scale. As a result, clean trends as a function of time and temperature were not observed.

Undercreep

It was anticipated that appreciable corrosion of the steel surface would occur under the lining around the holiday, particularly on the non-blast-cleaned surface where adhesion was marginal. Instead, little corrosion was observed around the holiday, even at elevated temperatures for extended immersion. The instances for which undercreep was observed are summarized in Table 6.5. The results illustrate the following:

- The epoxy lining system is not highly susceptible to undercreep. The instances of occurrence were limited and in most cases, required 8 months to initiate.
- Undercreep is more likely to occur at low DFT (most of the occurrences were on the 20-mil preparations).
- Undercreep is more likely to occur at higher temperatures (most instances were at 35 °C with only a few at 25 °C and none at 8 °C).
- Most instances of undercreep were on the non-blast-cleaned side, as anticipated.
- Undercreep susceptibility was not strongly related to adhesion (based on knife adhesion). Epoxy A and polyurethane linings had poor knife adhesion on the non-blast-cleaned side, but had only isolated instances of undercreep.

It was not feasible to determine the loss of lining adhesion around the holiday based on the amount of lining that could be pried away around the holiday. The lining broke away from the holiday in varying amounts, depending on the exact prying action, lining thickness, and lining mechanical strength. For comparison, the lining was also pried along the radial cuts, well away from the holiday. In general, the size of the chips and nature of the lining separation was similar at both the holiday and along the scribe, well removed from the holiday.

Underfilm Discoloration and Rusting

The knife adhesion tests and undercreep tests exposed areas of the steel surface. These areas were examined and rated for rusting using the pictorial standards in ASTM D610. Rusting was considerably easier to observe on the bright, blast-cleaned surface (side A) compared to the mill-scaled surface (side B). The instances of underfilm corrosion are summarized in Table 6.6.

Table 6.5
Summary of undercreep results

Lining	Temp, °C	Immersion time (months)	Panel side	No. of panels	Sectors	Undercreep
Epoxy A, 20 mils	25 °C	8	A	1	1	3 mm
Epoxy A, 20 mils	25 °C	8	B	1	2	1–2 mm
Epoxy A, 20 mils	35 °C	2	B	1	2	5 mm
Epoxy A, 20 mils	35 °C	8	B	2	2 -3	1–3 mm
Epoxy A, 20 mils	35 °C	8	A	1	6	2–3 mm
Epoxy B, 20 mils	25 °C	8	B	1	6	1–2 mm
Epoxy B, 20 mils	35 °C	2	B	1	6	1 mm
Epoxy B, 20 mils	35 °C	8	B	2	3	1–2 mm
Epoxy B, 40 mils	35 °C	4	B	1	6	1 mm
Epoxy B, 40 mils	35 °C	8	B	1	6	1, 2, 2, 5, 11, 5 mm for 6 sectors
Epoxy D	35 °C	8	B	1	6	6–9 mm
Polyurethane	35 °C	2	A	1	6	3–5 mm

Table 6.6
Summary of underfilm rusting results

Lining	Temp. °C	Immersion time	Panel side	No. of panels	Rust rating**
Epoxy A, 20 mils	25 °C	8 months	A	3	4P, 5P, 6P
Epoxy A, 20 mils	35 °C	1 month	A	4	6P
		8 months*	A	4	1P
		8 months	B	3	3G, 1P, 4G
Epoxy A, 40 mils	35 °C	4 months	A	2	4P
		8 months*	A	4	3P, 5P, 6P, 6P
		8 months	B	3	3G, 6G, 7G
Epoxy B, 20 mils	35 °C	2 months	B	2	6P
		8 months*	B	1	8G
Epoxy B, 40 mils	35 °C	8 months	B	1	8S
Epoxy D	8 °C	8 months	B	2	7G, 8G
Epoxy D	25 °C	2 months	B	4	7G
		8 months*	B	1	9G
		8 months	A	2	4G, 6G
Epoxy D	35 °C	2 months	B	4	7G, 7G, 8G, 8G
		8 months*	B	4	3G, 4G, 5G, 6G
Polyurethane	35 °C	2 months	A	1	1G

*After the initial observation, the rusting continued to grow until reaching this stage at 8 months.
**See Appendix D for definitions of rust rating values.

The results are summarized as follows:

- Significant amounts of rusting developed on the blast-cleaned surface under the epoxy A lining.
- Epoxy C was the only lining that did not develop any rusting (polyurethane showed only one isolated instance).
- Significant amounts of rusting developed on the non-blast-cleaned surface under the epoxy D lining.
- Based on the lining, DFT had an important effect, where the rusting was considerably worse under the 20-mil lining compared to the 40-mil lining.
- In general, the tendency for rust to develop increased with temperature, with most instances of rusting observed where the immersion temperature was 35 °C.

Blistering

Blistering was not observed on any of the panels, with the possible exception of tiny surface blisters on several isolated epoxy A panels. Microscopic examination indicated they may simply be overspray or a function of the surface finish. The absence of blistering may be related to the absence of salt contamination on the steel surface, and thus, the absence of osmotic effects, which commonly drive the formation of blisters and lining disbondment.

Rust Stain and Corrosion Product Precipitation at the Holiday

As the immersion time progressed, a rust stain started to develop on the surface of the lining at the holiday. Based on undercreep analysis, the stain resulted primarily from corrosion of the bare steel exposed at the holiday. The rust stain was observed first on the panels at 35 °C, and then progressively appeared at 25 °C and 8 °C as the immersion time approached 8 months. At the 8-month interval, a small mound of corrosion products ($\frac{1}{16}$ inch in size) was observed to protrude from the holiday of some panels at 35 °C. At 8 °C, the holiday generally tended to remain clean and bright.

Other Observations

The surface of the epoxy A lining was very tacky when the lining was applied, after 16 hours of cure at ambient temperature. The tackiness persisted throughout the immersion, and was considerably more noticeable at 8 °C than at the higher temperatures. Also, epoxy A remained relatively soft, based on cutting and prying with a knife. Finally, widely varying degrees of beaching were observed in immersion testing. The bleaching did not follow any discernible pattern related to lining application, DFT, or test conditions. However, it did increase with increasing immersion time. At the end of 8 months, the panel surface ranged from very little, if any, change in color or gloss, to a severe change in color and gloss where the lining surface was dull and almost white. The bleaching was limited to the surface and did not have any obvious relationship to performance.

Pull-Off Adhesion Results

The results are summarized in Figures 6.19 through 6.25. The results are described in the following paragraphs.

Epoxy B and polyurethane had the highest pull-off adhesion values of the five linings, with all pull-off values 2,300 pounds per square inch (psi) or greater. In many cases the adhesion exceeded the upper limit of 2,500 psi for the pull-off fixture. The pull-off values varied only slightly with immersion temperature and immersion time, but not in any systematic way. DFT did not have any significant effect on the adhesive strength of epoxy B.

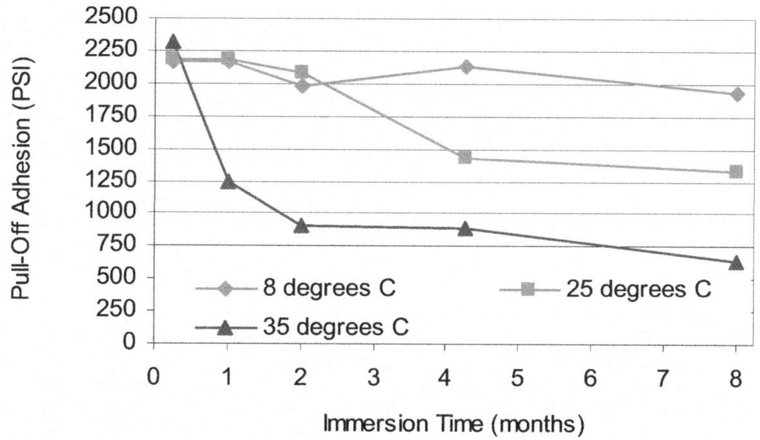

Figure 6.19 Pull-off adhesion results – epoxy A 20 mils

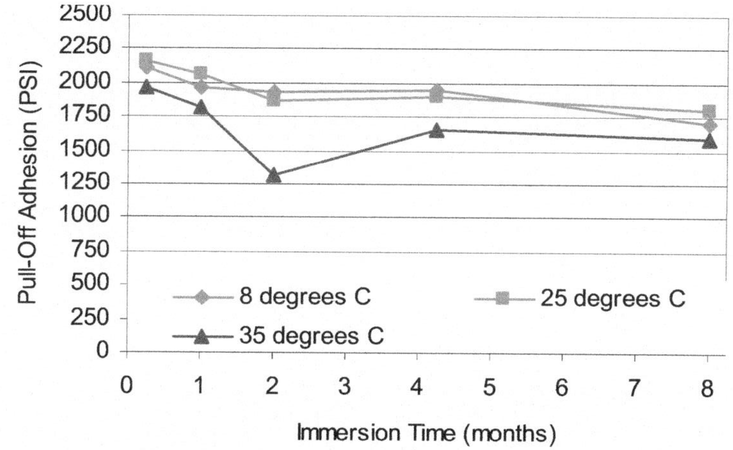

Figure 6.20 Pull-off adhesion results – epoxy A 40 mils

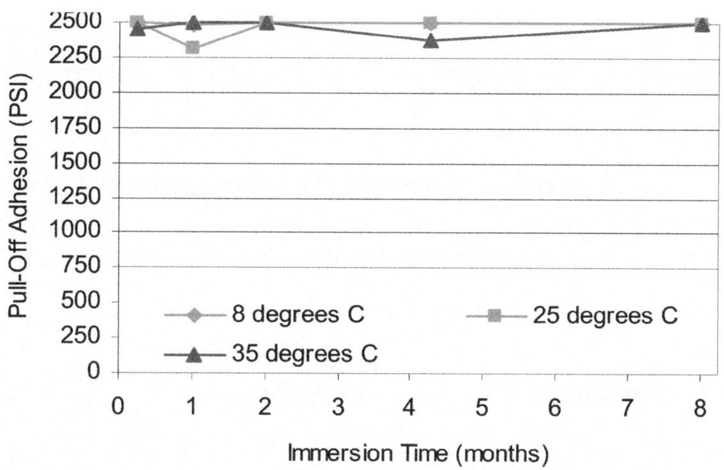

Figure 6.21 Pull-off adhesion results – epoxy B 20 mils

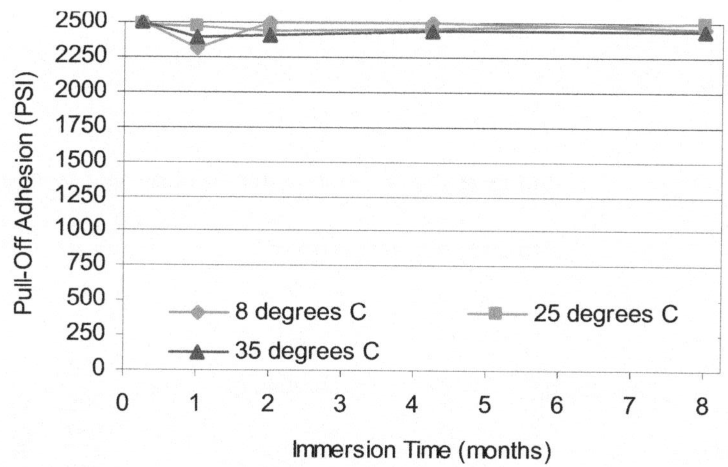

Figure 6.22 Pull-off adhesion results – epoxy B 40 mils

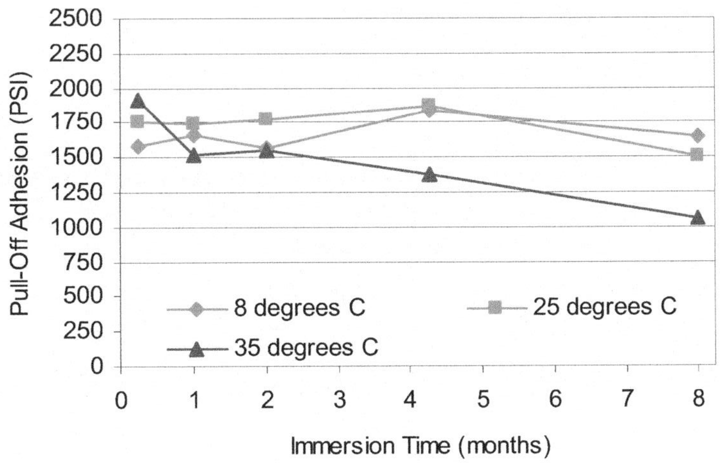

Figure 6.23 Pull-off adhesion results – epoxy C 40 mils

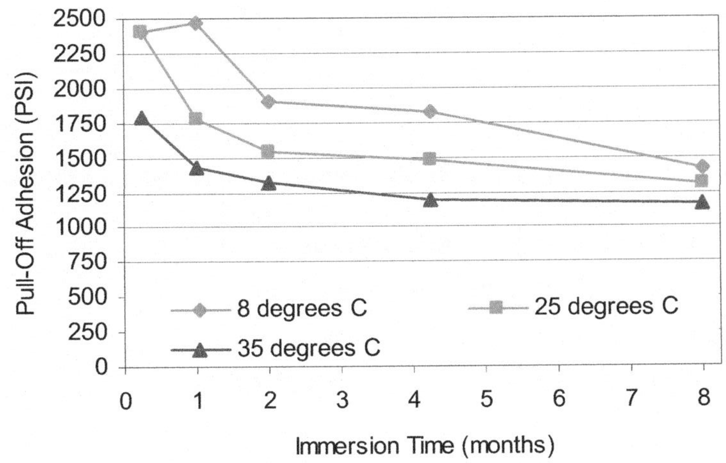

Figure 6.24 Pull-off adhesion results – epoxy D 40 mils

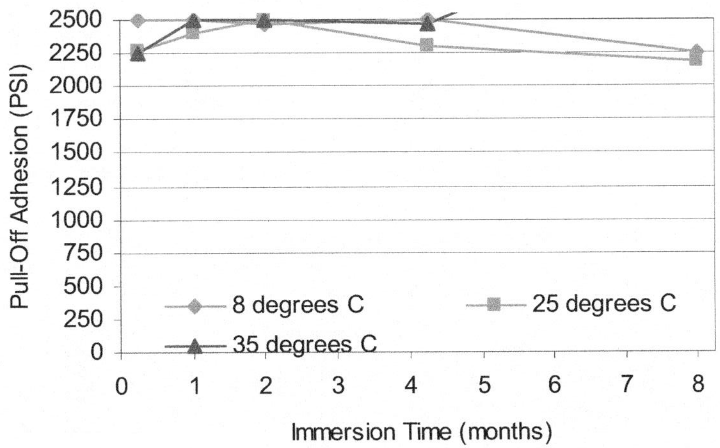

Figure 6.25 Pull-off adhesion results – polyurethane 40 mils

The pull-off values for the epoxy D lining decreased rapidly over the first 2 months of immersion (at all three temperatures) after which the values either stabilized (35 °C) or continued to decrease at a reduced rate (8 °C and 25 °C). The pull-off values decreased systematically with increasing temperature. After 8 months of immersion, the pull-off values were 1,417, 1,300, and 1,167 psi at 8 °C, 25 °C, and 35 °C, respectively.

The pull-off values of the epoxy A lining varied significantly, depending on the immersion temperature, immersion time, and DFT. After 8 months of immersion, the pull-off adhesion of the 40-mil film had pull-off values of 1,700 psi, 1,800 psi, and 1,600 psi, respectively, at 8 °C, 25 °C, and 35 °C. In contrast, the pull-off adhesion of the 20-mil film decreased dramatically with increasing immersion temperature and time. After 8 months of immersion, the pull-off values were 1,933 psi, 1,300 psi, and 633 psi at 8 °C, 25 °C, and 35 °C, respectively. Under some conditions, the pull-off adhesion appeared to be decreasing very slowly with increasing immersion time. It is generally accepted within the linings industry that pull-off adhesion should be a minimum of 700 psi for immersion service.

The pull-off values of the epoxy C lining were not reliable, due to intercoat delamination. Of the data obtained, there was no discernible trend as a function of immersion time at 8 °C and 25 °C; the pull-off values varied over a range of 1,500 to 1,850 psi over the 8-month duration of the immersion. At 35 °C there was a consistent downward trend in pull-off adhesion as a function of immersion time, dropping from an initial value of about 1,900 psig at 1 week to 1,070 psi after 8 months of immersion. The second application was carried out well within the recoat window of the product and therefore, the intercoat delamination was very unexpected.

Buchholz Hardness

The Buchholz hardness results are shown in Figures 6.26 through 6.32. The results are summarized in the following paragraphs.

Epoxy B and polyurethane were the hardest linings of the group, with an indentation length of about 1.0 mm. The hardness was not significantly affected by immersion temperature or immersion time. Only small random deviations from 1.0 mm occurred, which are attributed to the natural variation in the smoothness of the surface and flatness. There was no difference in hardness between the two DFTs of the epoxy B lining.

Epoxy C and epoxy D were somewhat similar in behavior. Epoxy C had an indentation length of 1.2 mm. In the case of the epoxy B lining, it showed little change in hardness from hydration and immersion temperature. The epoxy D lining appeared to continue to cure, reaching full hardness after about 2 months of immersion. The hardness was greatest at 8 °C.

The epoxy A lining hardness varied with immersion temperature, immersion time, and DFT. The epoxy A 40-mil preparation continued to become harder with increasing immersion time, and presumably continued to cure with time. At 8 °C, 8 months was required to reach what appeared to be near to full hardness. Between 2 and 4 months was required to reach almost full hardness at 35 and 25 °C, respectively. The hardness after 8 months of immersion was an indentation length of 1.4 mm.

Epoxy A (20-mil preparation) at 8 °C started out very soft and continued to become harder with immersion time. After 8 months of immersion, the hardness was still on an upward trend. The 35 °C preparation also hardened with immersion time, reaching maximum hardness within 4 months. The 8 °C and 35 °C preparations reached indentation lengths of 1.6 and 1.4 mm. In contrast, the 25 °C preparation increased in hardness during the first month of immersion, and then gradually decreased in hardness. After about 3 months of immersion, the hardness of the 25 °C preparation passed below that of the 8 °C preparation.

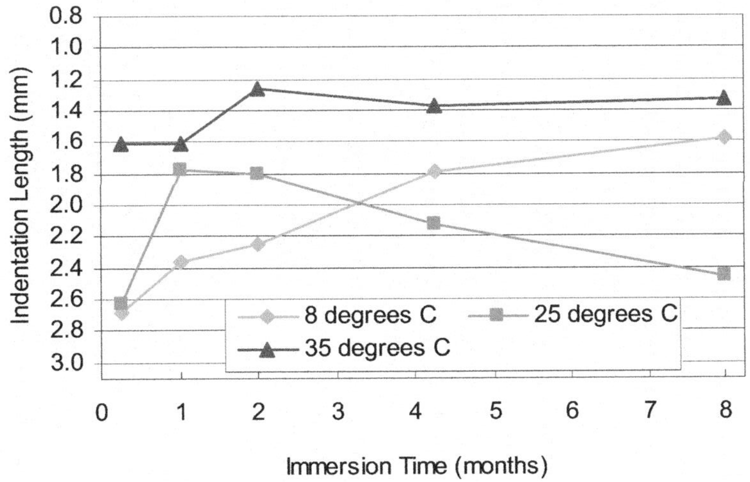

Figure 6.26 Buchholz hardness results – epoxy A 20 mils

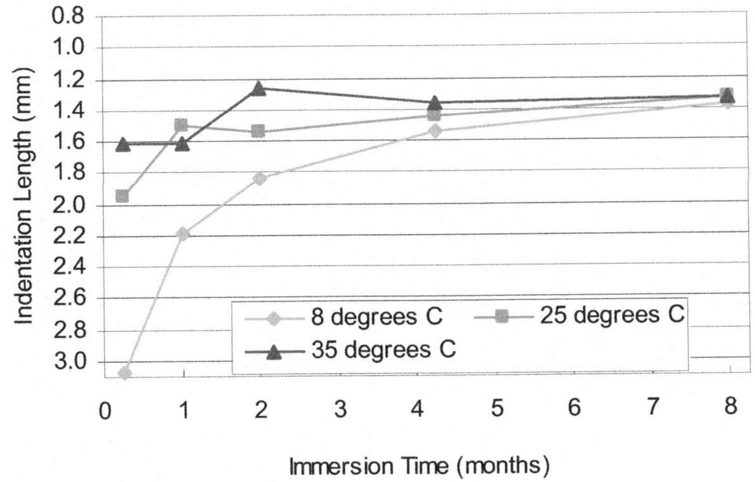

Figure 6.27 Buchholz hardness results – epoxy A 40 mils

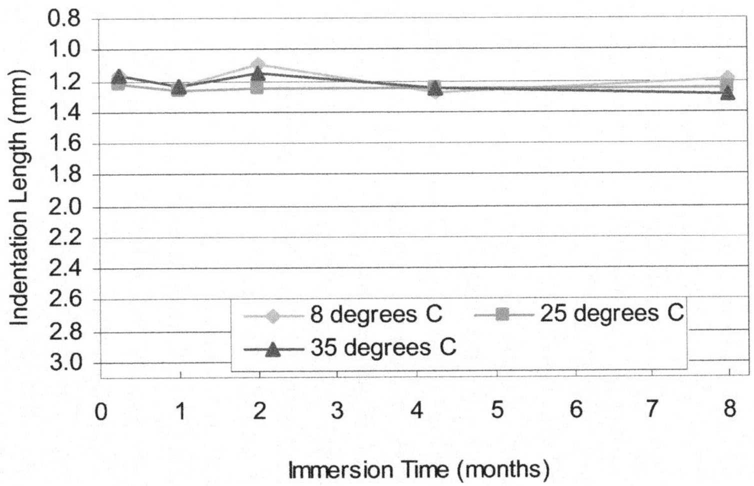

Figure 6.28 Buchholz hardness results – epoxy B 20 mils

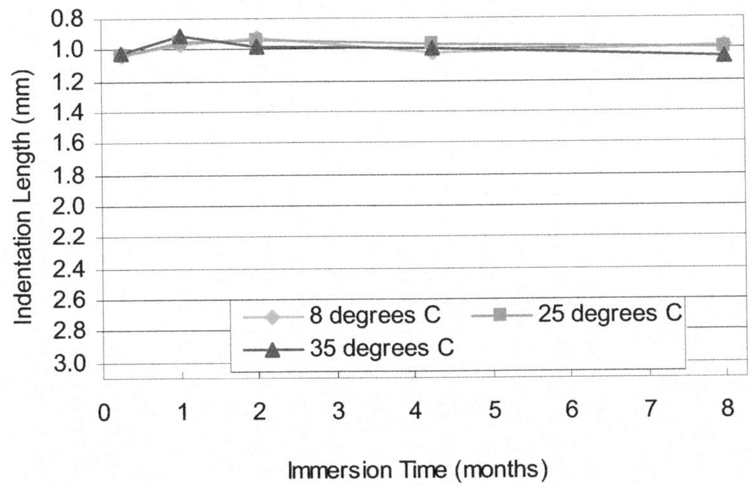

Figure 6.29 Buchholz hardness results – epoxy B 40 mils

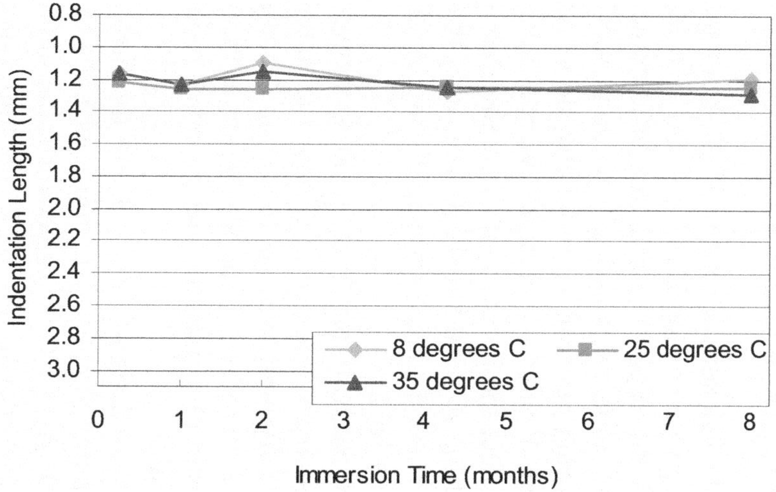

Figure 6.30 Buchholz hardness results – epoxy C 40 mils

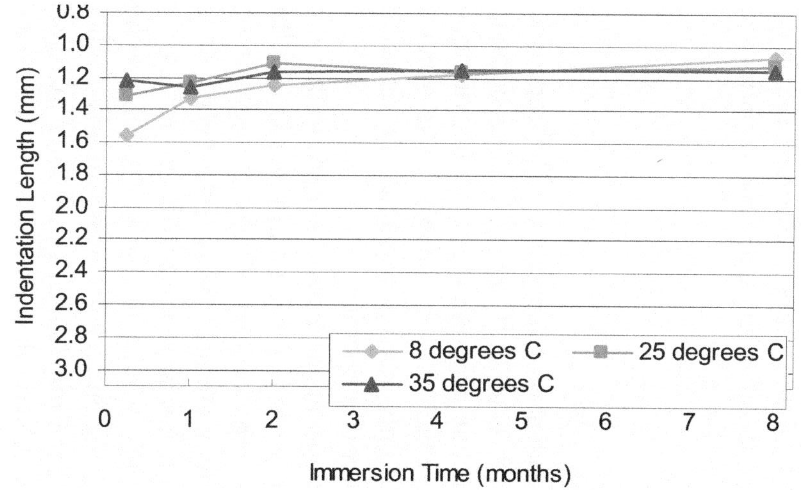

Figure 6.31 Buchholz hardness results – epoxy D 40 mils

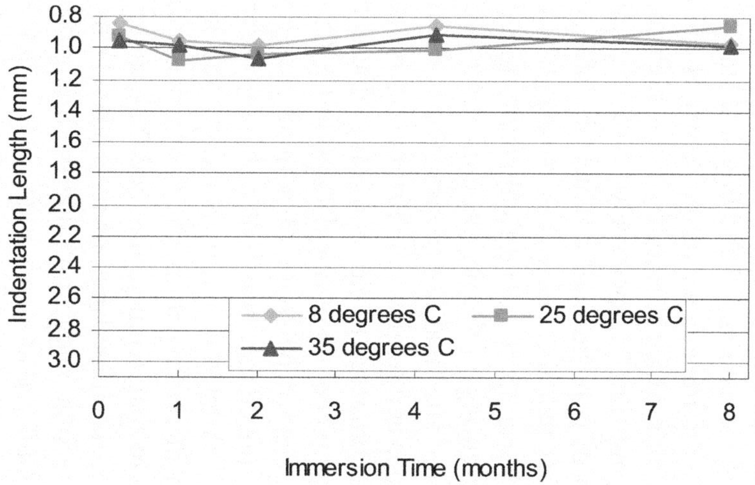

Figure 6.32 Buchholz hardness results – polyurethane 40 mils

The epoxy A lining was significantly softer than any of the other linings for most of the immersion period. This is consistent with the sticky feel of this lining. Further, the larger variation between replicate data and the varied behavior is consistent with the variations in bleaching, which suggests some variation in lining properties.

EIS Results

The EIS results are presented in Figures 6.33 through 6.39.

The impedance of epoxy A (20 mils) was highly dependent on temperature. The impedance decreased significantly as the temperature was increased from 8 °C to 25 °C to 35 °C. This behavior indicates that as the temperature is increased, the polymer structure becomes more mobile with more free volume (empty space between the polymer molecules), which, in turn, enables more water to diffuse through the lining (i.e., as the temperature is increased, the barrier properties are reduced). The impedance at 35 °C was log Z = 8.5, indicative of good barrier properties.

The other significant observation was that the impedance increases during the first several months of immersion. This behavior suggests that the lining was continuing to cure for the first several months, resulting in a more highly cross-linked structure with lower permeability. After 4 months the impedances at all temperatures were relatively stable, indicating that no further chemical or physical change was occurring in the barrier properties.

The impedance of the epoxy A 40-mil lining was very similar to that at 20 mils, with a few differences. At each of the respective temperatures, the impedance was slightly higher, undoubtedly as a result of the higher DFT. Theoretically, the impedance of a lining increases linearly with thickness. Also, at 8 °C and 25 °C, the impedance decreased during the first 2 months of immersion, suggesting that the lining was taking up water or undergoing some physical or chemical changes that resulted in increased permeability. However, after these initial adjustments the impedance was stable for the duration of the immersion, suggesting that the film would have long-term chemical and physical stability.

At 8 °C, the impedance of the epoxy B lining climbed steadily during the first 4 months of immersion, after which it stabilized at a high impedance value of log Z of 10.5 and 10.8 for the 20- and 40-mil films, respectively. At 23 °C and 35 °C, the impedance also increased during the first month of exposure, after which it stabilized at the same value as the 8 °C immersion. The data suggest that the lining was curing, where the curing rate was considerably slower at 8 °C than at 23 and 35 °C. The final stable impedances at all three temperatures were similar, suggesting that curing in the presence of water does not affect the integrity of the lining that is finally achieved. The final impedance was in the range of log Z of 10.5 to 10.8, indicating that epoxy B develops and retains excellent barrier properties. The impedance was slightly higher for the 40-mil film compared to the 20-mil film due to the higher DFT.

With the exception of one data point (8 °C, 1-month interval), the epoxy C lining showed the classic impedance behavior of a lining in an immersion environment. During the initial immersion, the impedance was either stable or dropped to varying degrees as the lining became fully hydrated. After the initial hydration, the impedance stabilized. The stable impedances decreased with increasing immersion, which is characteristic of a polymer that becomes more mobile with increasing temperature, and thus, more permeable.

At 8 °C, the impedance of the epoxy D lining increased significantly during the first 2 months of immersion, suggesting that the lining was continuing to cure. The impedance at 8 °C subsequently stabilized at a very high value of log Z = 11.0. At 25 °C the impedance was stable during the first month, after which it started a steady, but slow decline. At 35 °C, the impedance of the lining dropped quickly over the first month from a very high value (log Z = 10.6) to a log

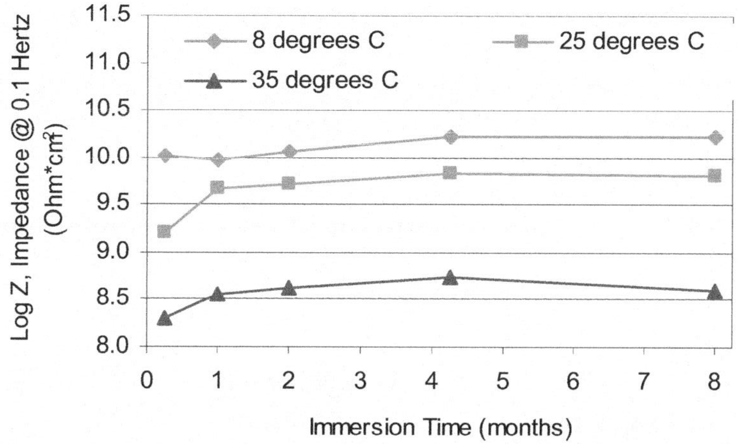

Figure 6.33 EIS test results – epoxy A 20 mils

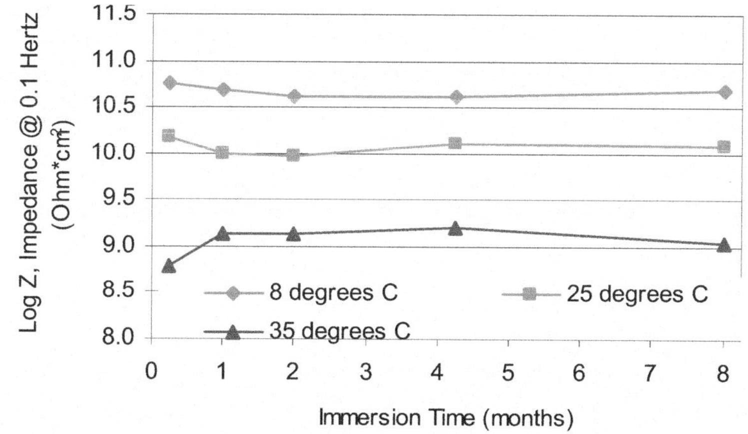

Figure 6.34 EIS test results – epoxy A 40 mils

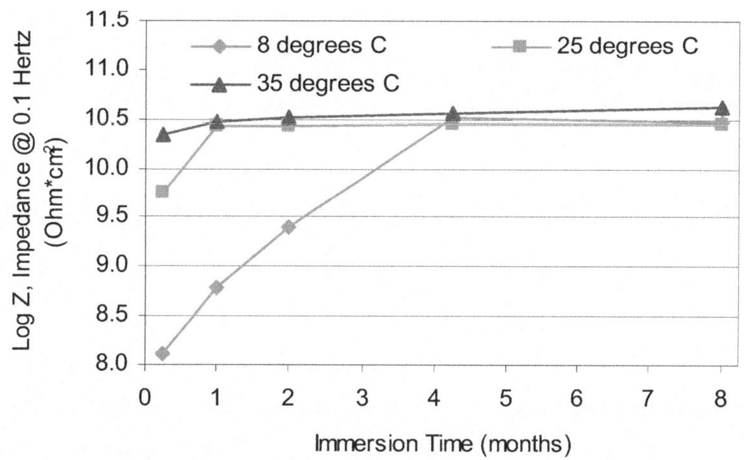

Figure 6.35 EIS test results – epoxy B 20 mils

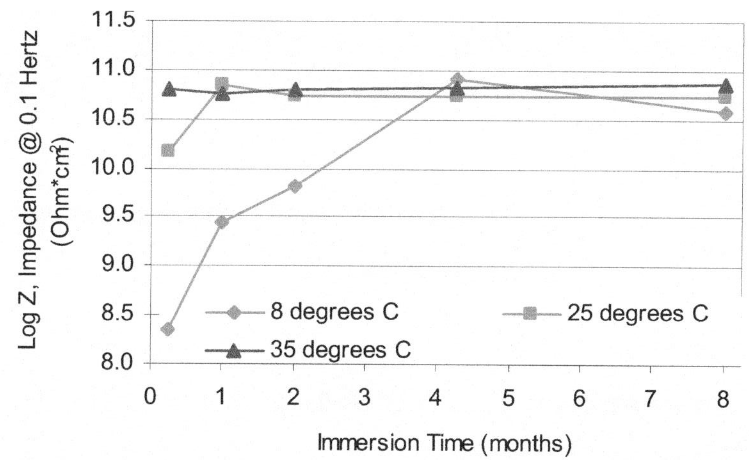

Figure 6.36 EIS test results – epoxy B 40 mils

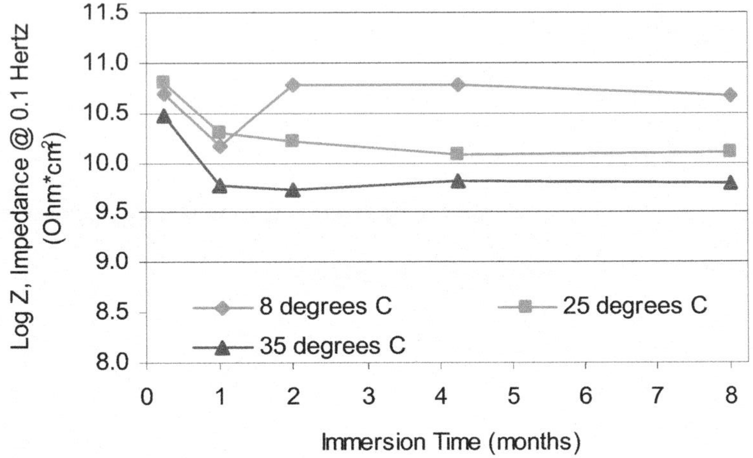

Figure 6.37 EIS test results – epoxy C 40 mils

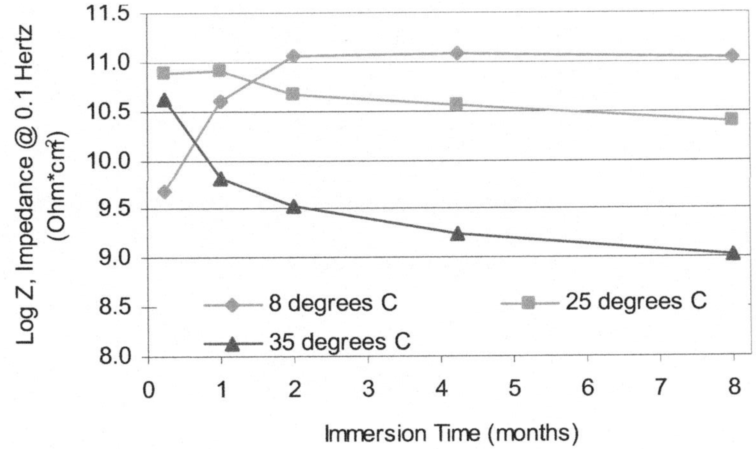

Figure 6.38 EIS test results – epoxy D 40 mils

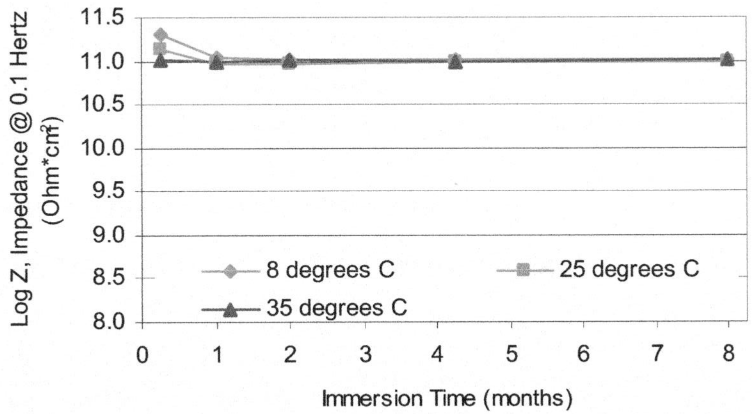

Figure 6.39 EIS test results – polyurethane 40 mils

Z of 9.8, after which it continued to steadily decline at a slower rate. The epoxy D lining showed a decrease in impedance with immersion time (at 25 °C and 35 °C).

The polyurethane lining was held at room temperature for an extended period of time during shipping, prior to being placed into immersion. The extended duration at room temperature was unlikely to have had a significant effect on the results because this coating was fully hardened within 10 minutes of application.

This lining showed the classic behavior of a fast-curing lining with outstanding barrier properties in immersion service. The impedance dropped slightly during the initial immersion as the lining hydrated, after which the impedance was stable with no further change at 8 °C, 25 °C, and 35 °C as a function of time. The stable impedance was log Z = 11 at all three temperatures, indicative of a lining with exceedingly good barrier properties.

Lining Erosive Wear (ASTM G6)

The loss of DFT and weight for each of the linings is shown in Table 6.7. The data are summarized in Figure 6.40, which shows the weight losses expressed in terms of rotational cycles per mil and per gram of lining loss. This format provides the relative performance on a scale between 1 and 10. Photos of the tested samples are shown in Figure 6.41.

The worn areas were uniformly worn, without any chipping. At the completion of the wear test, the 20-mil preparations of epoxy A and epoxy B were almost completely worn away along a thin strip along the length of the tube where the wear was at a maximum. The reference points at which the loss of lining was measured were immediately adjacent to this area of maximum wear. Therefore, the data in Table 6.7 represent relative wear rates of the linings. A summary of the results and visual observations for the various linings are discussed in the following paragraphs.

Epoxy A and epoxy C had the highest rates of wear, with average DFT losses of 11.2 and 9.4 mils, respectively. The worn area of the epoxy A surface had lost its gloss and was roughened.

Epoxy B had the next highest wear rate, with average DFT losses of 8.1 mils for the 20- and 40-mil preparations. Based on the standard deviation, there was no significant difference between the 20- and 40-mil preparations.

Epoxy D and polyurethane had the lowest amounts of wear, with an average DFT loss of 4.5 and 5.3 mils, respectively. The worn areas showed only a slight loss of gloss, and under the microscope, were observed to be slightly roughened. Polyurethane would be anticipated to have a lower rate of wear than epoxy.

The weight changes of the various linings correlated very well with the DFT changes, except for epoxy D. Inspection showed that more abrasive media was trapped within the worn surface of the epoxy D lining, reducing its apparent weight loss compared to the DFT loss.

The value of these results lies in the relative wear resistances of the five linings, rather than the absolute amount of wear. The conditions of the test were very aggressive in comparison to potable water. To summarize, the linings with the poorest wear resistance, wore about twice as fast as the best linings. However, it should be noted that although abrasion and wear have been reported for epoxy linings, it is rarely a cause of failure.

SUMMARY OF RESULTS

The laboratory work revealed the following characteristics of the epoxy and polyurethane lining systems.

Table 6.7
Summary of wear results

		DFT loss (mils)											Weight loss		
Coating	ID#	P1	P2	P3	P4	P5	Min.	Max.	St. dev.	Avg.	Avg. cycles/mil (*10,000)	Weight loss (g)	Avg. loss (g)	Cycles/ gram (*100,000)	Avg. cycles/ gram (*100,000)
Epoxy A 20 mils	180 181	11.3 11.2	11.5 10.8	11.7 10.8	9.5 10.6	11.0 11.8	9.5	11.8	0.7	11.0	3.9	1.85 1.88	1.9	2.3 2.3	2.3
Epoxy A 40 mils	280 281	9.5 13.2	10.3 11.9	9.6 12.5	11.3 11.0	11.2 12.4	9.5	12.5	1.2	11.3	3.8	2.03 2.08	2.1	2.1 2.1	2.1
Epoxy B 20 mils	380 381	9.0 8.4	7.7 8.8	7.0 9.0	8.5 8.2	8.0 9.8	7.0	9.8	0.8	8.4	5.1	1.65 1.59	1.6	2.6 2.7	2.7
Epoxy B 40 mils	480 481	8.2 8.4	7.6 7.2	7.1 7.5	7.7 7.0	8.0 9.0	7.0	9.0	0.6	7.8	5.6	1.45 1.54	1.5	3.0 2.8	2.9
Epoxy C	580 581	12.4 8.5	10.4 8.9	9.9 7.3	10.3 8.1	10.4 7.6	7.3	10.4	1.6	9.4	4.6	1.78 1.72	1.7	2.4 2.5	2.5
Epoxy D	680 681	5.2 5.1	4.6 4.8	4.3 3.1	4.2 3.2	5.3 5.2	3.1	5.3	0.8	4.5	9.7	0.46 0.50	0.5	9.4 8.7	9.1
Polyurethane	780 781	5.2 5.6	4.3 5.7	3.7 6.2	4.4 6.2	4.4 7.2	3.7	7.2	1.1	5.3	8.2	1.37 1.85	1.6	3.2 2.4	2.8

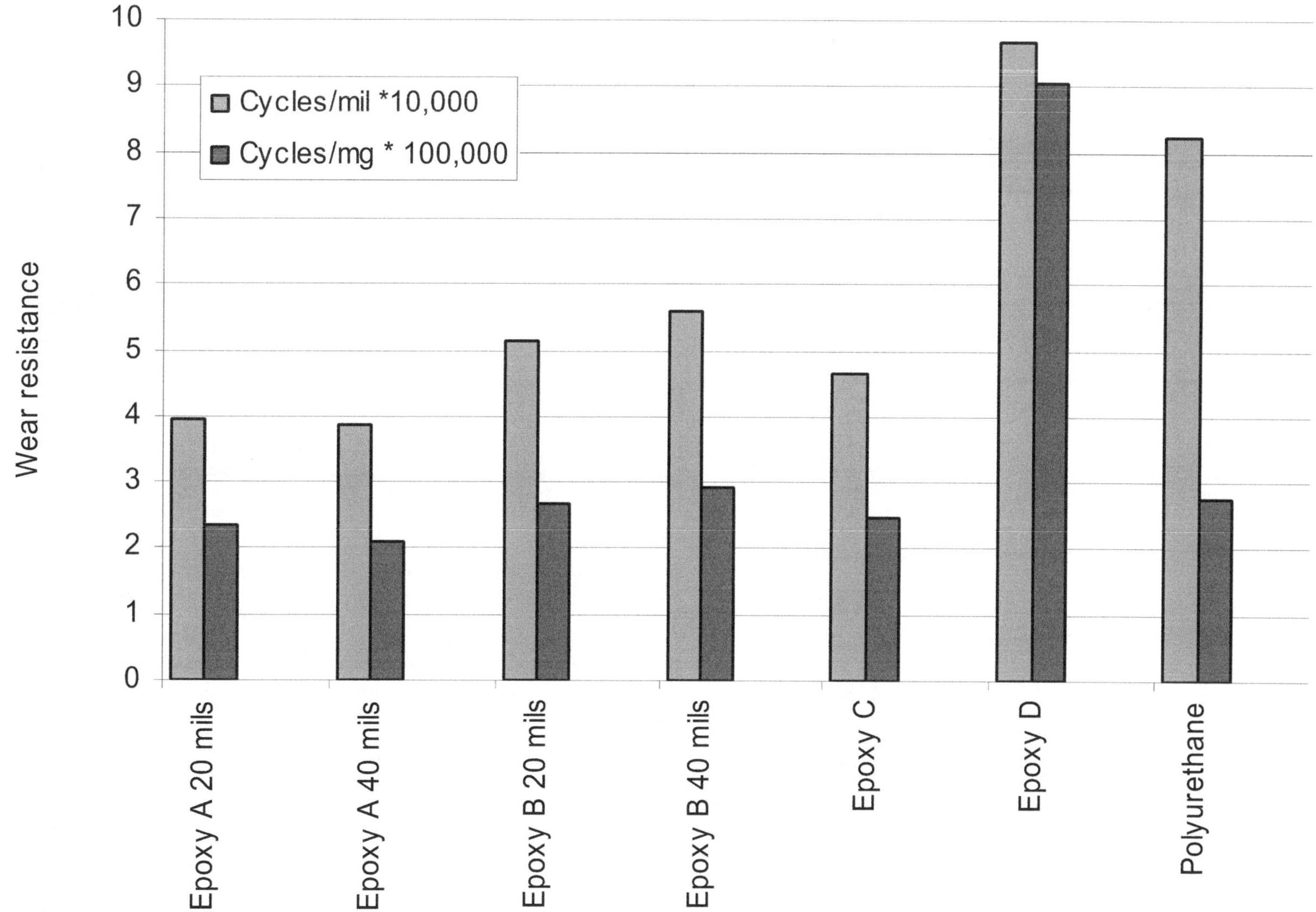

Figure 6.40 Wear test results

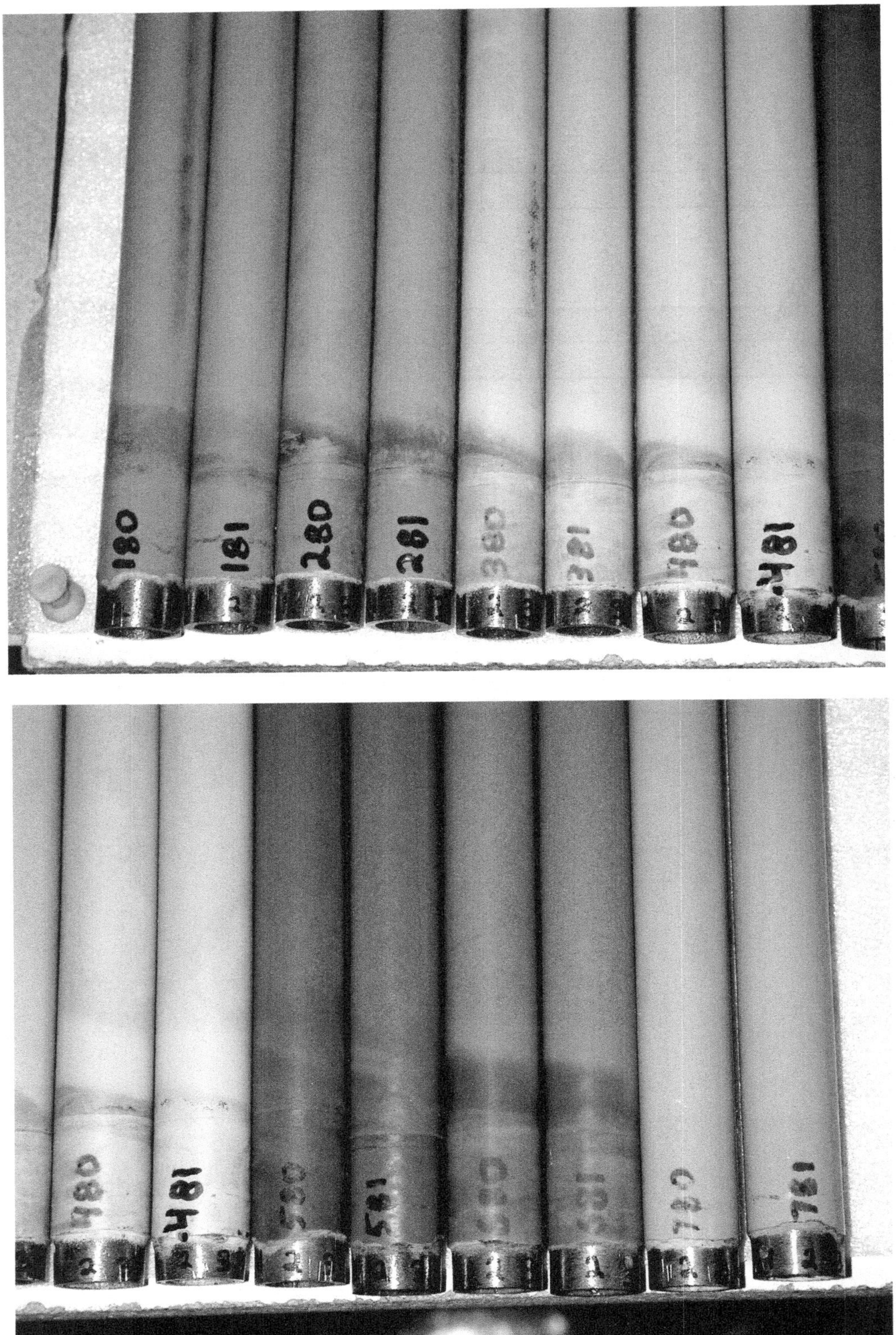

Top: Epoxy A (blue) Epoxy B (grey)
Bottom: Epoxy C (red) Epoxy D (green) Polyurethane (grey)

Figure 6.41 Wear test samples

Epoxy A

Epoxy B

Epoxy C

Epoxy D

Polyurethane

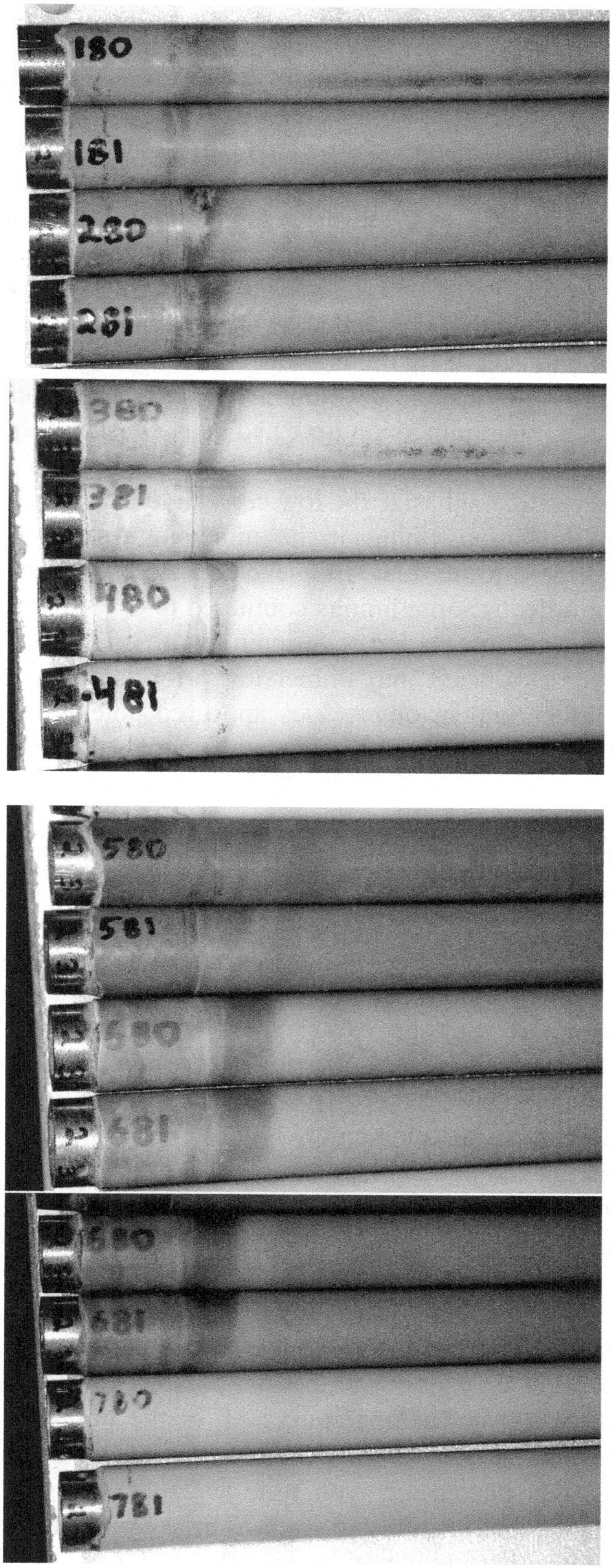

Figure 6.41 Wear test samples (continued)

The immersion environment, consisting of aerated distilled water at temperatures of 8 °C, 25 °C, and 35 °C, had unexpectedly low corrosivity. Exposed carbon steel at holidays experienced very limited corrosion, particularly at 8 °C, which is close to many soil temperatures. Undercreep developed under some linings, but it was limited in extent, and related primarily to low film build-up and elevated temperature. Crevice corrosion (accelerated corrosion under the lining near the holiday) was not a mechanism of corrosion in this system. Underfilm rusting was similarly limited, and related primarily to low film build-up at elevated temperature and linings with higher permeability. These observations were consistent with those for field samples and previous investigations (Warren, 2004).

Deterioration of the linings and development of undercreep and underfilm rusting would have been more extensive if the immersion time had been considerably longer. The data suggest, however, that many multiples of the current 8-month immersion time would have been required. Therefore, the results serve to emphasize the low corrosivity of the potable water system.

The behavior of the various linings in the immersion test was considerably more complex than expected. This appeared to be related to several different processes that were occurring in some linings and not in others. Some linings continued to cure after being placed in immersion, where the extent and rate of cure varied with temperature. The role of water is unclear in the performance of the epoxies placed into immersion prior to full cure. In some cases, water appeared to have no effect, and in other cases appeared potentially detrimental. Some of the linings became much more permeable to water with increasing temperature. Finally, chemical and physical changes associated with lining deterioration appeared to be occurring in some linings, based on decreasing impedance with time.

Epoxy C retained good adhesion on both blast-cleaned and non-blast-cleaned surfaces. It had high and stable hardness, good barrier properties, good pull-off adhesion, and no instances of undercreep or underfilm corrosion.

Epoxy B had retained relatively good adhesion, had excellent barrier properties, good hardness, and only a limited amount of underfilm rusting and undercreep. Polyurethane had somewhat low adhesion, as may be expected for polyurethane in an immersion service. However, it had outstanding hardness, barrier properties, and showed little tendency for underfilm rusting and undercreep.

Epoxy D had low adhesion, moderately low barrier properties, and development of some degree of underfilm rusting and undercreep.

Epoxy A data indicate that it had good adhesion. It had low hardness, and moderate barrier properties, which decreased with increasing temperature. It also showed undercreep and underfilm corrosion.

Of the various parameters that were monitored in linings, impedance was the most useful one in terms of estimating lining life expectancy. The impedance data indicated that the lining deterioration processes were very slow and that the potable water environment is not aggressive even with acceleration at elevated temperature. All of the linings generally had very good to excellent barrier properties with log Z greater than 10. Epoxy D was the only lining that had impedance that was slowly decreasing with time at 25 °C and 35 °C, indicative of possible deterioration and increasing permeability to water. These observations qualitatively indicate that the epoxy and polyurethane linings would be expected to have a very long life expectancy, i.e., many multiples of the 8-month duration of the immersion test.

Adhesion of the seven linings was measured in two different ways: knife adhesion and pull-off adhesion. The two methods provided very different results. The reason for the difference is that the two methods use very different mechanical actions on the lining. In the knife adhesion test, adhesion is evaluated based on forcing a utility knife between the lining and metal, followed by a prying action to remove the lining from the surface. In contrast, pull-off adhesion consists of applying a force normal to the lining surface, where both cohesive and adhesive forces must be overcome almost instantaneously over a much larger area (i.e., 0.75-inch diameter dolly compared to the tip of a utility knife).

Other items of note are as follows.

- Blistering was not observed on any of the laboratory samples, which is consistent with field sample observations where blistering was only rarely observed.
- Corrosion of the steel surface under the lining was related to poor barrier properties of the lining, which was further aggravated by low film buildup and elevated temperature.
- The laboratory data for plain carbon steel were comparable to field data for potable water lines, which are constructed of ductile iron or cast iron. The similarity in observations between the laboratory and the field samples indicates that the corrosion processes and mechanism are similar for the two surfaces.
- It was not possible to accelerate the test further by increasing the immersion temperature above 35 °C because the glass transition temperature of most of the epoxies is only slightly above 35 °C. Above the glass transition temperature, the mechanism of deterioration and corrosion would be expected to change due to the large increase in the flexibility and mobility of the polymer.

CHAPTER 7
ESTIMATION OF USEFUL LIFE OF EPOXY LININGS

INTRODUCTION

Estimation of the useful life of an epoxy liner is one of the objectives of this study, although it is very difficult to quantify.

First, it is necessary to define the term useful life. From a utility perspective, as per discussions in a workshop conducted as a part of this study, the end of the useful life of a liner is considered when water quality issues return. When complaints from customers become frequent, the lining has reached the end of its useful life. However, data on water quality problems and customer complaints are not readily available to estimate useful life.

The most important parameter for proper functioning of an epoxy liner is its barrier property measured by impedance (using EIS). It is also noted in the coatings industry that an EIS value of 6 is considered to represent the end of the useful life of a liner. In this study, an approach to estimate the useful life of a liner is based on comparing EIS values of historical pipe liners with EIS values of accelerated laboratory testing of coatings.

LITERATURE ON ACCELERATED AGING TESTS

Accelerated aging tests represent one method of estimating the useful life of epoxy coatings. Some examples of accelerated aging tests of epoxy liners have been described by others (S.J. Canfer, 1990; D. Walker and S. Guan, 1997; SSPC, 1990).

An accelerated aging test, reported to be equivalent to 100 years of use in potable water service, was performed on Mercol Geopox GX014 epoxy resin (Canfer, 1990). Quadrants of cast iron pipe sections that had been routinely lined in situ were used as test specimens. The specimens were placed in a solution of deionized water with 100 mg/L chloride (as sodium chloride) and 100 mg/L sulfate (as sodium sulfate) at a temperature of 50 °C. The edges of the lining were undermined to encourage corrosion of the cast iron in specific areas. An acceleration factor of 12 for cast iron corrosion was reported. The test solution was renewed every 14 days and the test was conducted for a period of 8 years and 4 months. At the end of the test period the cast iron pipe was visually examined. The cast iron pipe was corroded but the epoxy was still adhering "satisfactorily" to the pipe. The epoxy liner has lost some color, but there was no significant embrittlement and the liner could still be flexed by hand. The report also contains technical specifications for Geopox GX014 including material test results for salt spray (ASTM B117); physical abrasion (Tabor); water erosion, tensile strength, flexural strength (DIN 53452); and modulus of elasticity, shrinkage (cure), hardness, mix ratio tolerance, and blister formation.

Walker and Guan (1997) reported test results comparing 5 categories including 13 different ANSI/NSF 61 approved lining systems used for steel pipelines in potable water service. The five categories of lining systems were solvent-amine-based epoxy (6), 100% solids epoxy (2), 100% solids elastomeric polyurethane (2), 100% solids rigid polyurethane (2), and cement-mortar (1). The liners were tested for 1,000 hours for salt spray resistance (ASTM B117) and cathodic disbondment (ASTM G95, 3% NaCl, 1.5 volts, 30 days, 73 °F). The report indicated that both tests could be used to predict the lining's ability to resist undercutting from the corrosion process, but the cathodic disbondment test was more consistent in predicting actual liner performance. The liners were also exposed to four different chemical solutions (sodium hydroxide, 20%; sodium chloride, 3%; sulfuric acid, 3%; and gasoline) for 1,000 hours in chemical resistance tests. All liners passed the chemical resistance tests for all four solutions.

No acceleration factors were calculated for the test conditions and the useful life of the coatings was not estimated.

The Steel Structures Painting Council (SSPC, 1990) performed five separate accelerated aging tests on two coatings (an epoxy material and a marine alkyd coating). The coatings were also subjected to two exterior exposure conditions. The five accelerated aging tests were:

1. Immersion in deionized water at variable temperatures
2. Immersion in tap water at high temperature
3. Immersion in deionized water under various pressure heads
4. Condensation cabinet at various temperatures
5. Acidified salt spray

The deterioration of the coating was measured by counting defects using a grid overlaid on the test panel. The report concluded that immersion in deionized water is the most aggressive to coating materials due to osmotic pressure effects, while salt solutions are more corrosive to steel. The deterioration of the coating was accelerated at elevated temperatures, and pressure had a small, but measurable, impact on coating deterioration.

METHODOLOGY TO ESTIMATE USEFUL LIFE OF EPOXY COATINGS

In this study a methodology for estimating the useful life of the coatings was developed using the EIS measurements taken from the immersion panels and the samples of historical epoxy-lined pipes that were submitted by participating water utilities for analysis. The data obtained from the immersion samples and the pipe samples were compared for similar epoxy materials. EIS was the only test that could be conducted on both test panels and pipe samples. The EIS test provided an objective quantitative measurement of the barrier properties of the epoxy material. The EIS measurements from the test panels (EIS versus immersion time) were plotted on the same graph as the historical pipe samples (Figure 7.1).

The four epoxy materials tested in the immersion test were Nitoline WP, Copon Hycote 162PWX, Geopox GX014, and ELC 257/91. The historical pipe samples were lined with the following materials:

- Geopox GX014 – 6 pipe samples
- ELC 257/91 – 3 pipe samples
- Nitoline WP – 1 pipe sample
- Copon Hycote 162PWX – 1 pipe sample
- Hunting Industrial – 2 pipe samples
- Nitoline WP2 – 1 pipe sample
- ELC 172/90 – 2 pipe samples
- Geopox GX012 – 1 pipe sample

The best comparison between laboratory and field data could be made for Geopox GX014 because six of the pipe samples were lined with GX014 epoxy. The EIS measurements for the immersion samples were plotted on the same graph with the pipe samples having the Geopox GX014 liners.

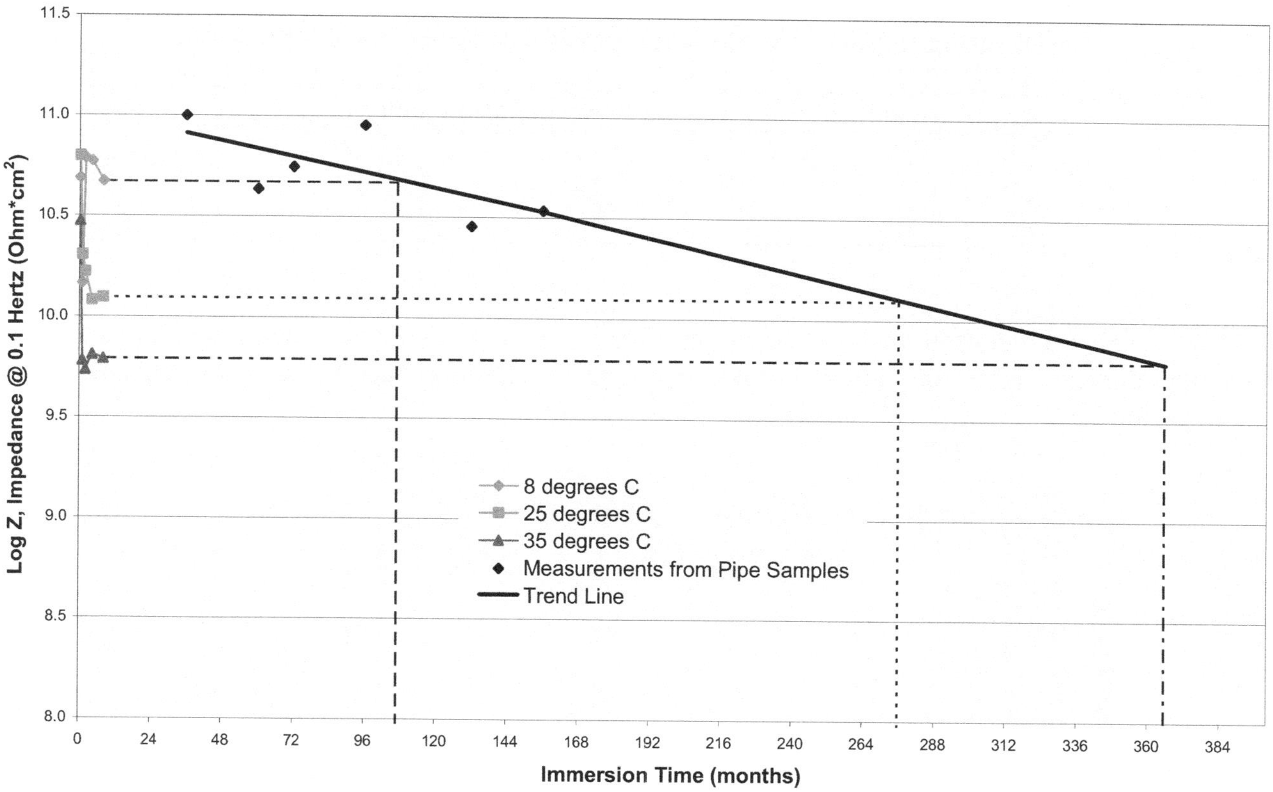

Figure 7.1 EIS versus in-service time

It should be noted that the EIS measurements were taken from locations on the pipe lining that were free of holidays. Therefore, the EIS measurements from the pipe samples are representative of the condition of the epoxy material and do not include the effects of holidays and other imperfections in the lining materials. The EIS measurements from the pipe samples lined with Geopox GX014 show a decreasing trend. A regression line showing the decreasing EIS trend with age of the field pipe samples is shown in Figure 7.1. The EIS measurements of the laboratory immersion panel coatings for Geopox GX014 at three different exposure temperatures of 8 °C, 25 °C, and 35 °C are also shown in Figure 7.1. The 8-month EIS measurements from the laboratory Geopox GX014 test panels were compared to the EIS regression line for the field pipe samples with Geopox GX014 liners to estimate the equivalent "field age" of the test panels. The ratio of "field age" to exposure time represents an acceleration factor of the accelerated aging tests performed during this study. In this way acceleration factors were calculated for the three exposure temperatures. The calculated acceleration factors are listed in Table 7.1.

Table 7.1
Acceleration factors at various temperatures for Geopox GX014 liners

Exposure temperature, °C	Acceleration factors
8	13
25	34
35	46

The acceleration factors shown in Table 7.1 show a linear relationship to exposure temperature as shown in Figure 7.2. As expected, acceleration factors increased with response temperature.

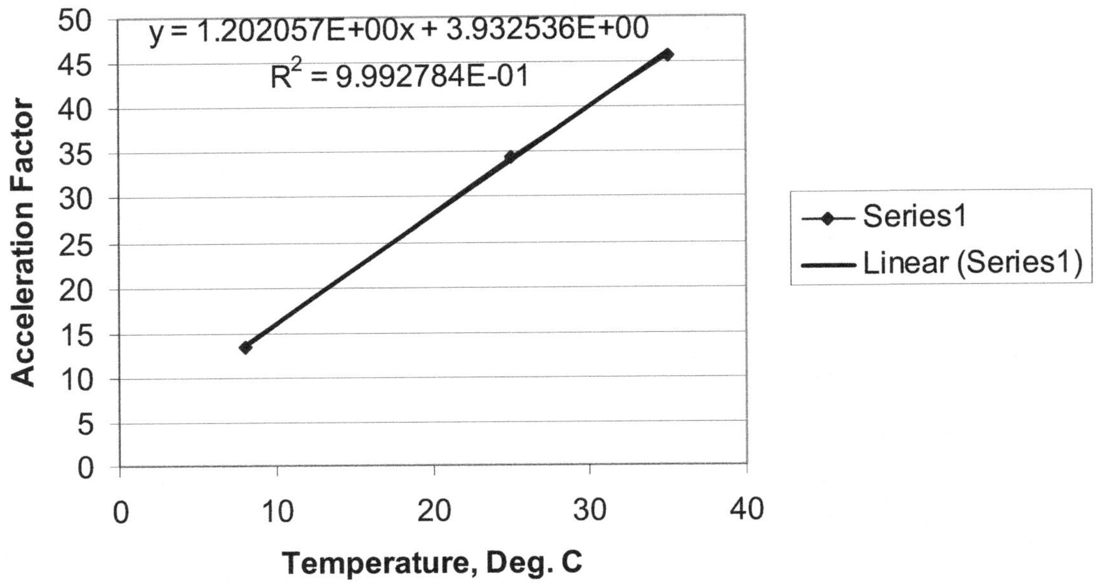

Figure 7.2 Acceleration factor versus exposure temperature

DETERMINATION OF USEFUL LIFE OF EPOXY COATINGS

In the coatings industry an EIS value of 6 (log Z) is considered to represent the end of the useful life of a coating and an EIS value of 4 represents total failure of the coating where active corrosion of the underlying metal can occur (Gray and Appleman, 2003).

The regression line for the Geopox GX014-lined pipe samples was extrapolated to EIS values of 6 and 4. The useful life of the Geopox GX014 liners was 123 years at an EIS value of 6. An EIS value of 4 would be achieved in 172 years. However, it should be noted that these values of useful lives are based only on the liner barrier properties and represent holiday-free liners. Any defects in the liners are expected to decrease the useful life of the liners. Assuming a safety factor of 2 to 3, the useful life of an epoxy liner can be estimated to be between 41 and 61 years. The value of the safety factor to be considered depends on the QC and the specifications of the lining construction followed during construction. In order to estimate the useful life of an epoxy lining an assumption of a safety factor should be made, the value of which depends on the quality of the lining construction. A higher value of the safety factor for low QC of a lining construction, and a lower value under high QC may be assumed.

It appears from this analysis that the useful life of an epoxy liner can be estimated to be in the range of about 40 to 60 years. Because epoxy lining, in general, is provided to rehabilitate old cast iron pipes, it does not make sense to put in a liner with a significantly higher life than the remaining life of the host pipe. It is expected that the useful life of a water main can be extended by 40 to 60 years by rehabilitating the main using epoxy lining.

CHAPTER 8
FINDINGS AND RECOMMENDATIONS

For old water mains in good structural condition, cleaning and lining is the most economical option for rehabilitation of these pipes. Many different cleaning and lining techniques are available. The effectiveness and life of these liners vary, depending on the physical properties of the liner material, the QC of liner construction, and the corrosivity of the water. If the proper liner is not used, the life of a rehabilitated pipe may be reduced significantly. In North America, most of the water main rehabilitation is conducted using cement mortar lining. Cement mortar lining was found to be ineffective as a long-term solution when water is relatively soft and aggressive. As an alternative to cement mortar lining, epoxy lining has been found to have high durability and higher hydraulic capacity. In the United Kingdom epoxy lining is predominantly used for small diameter water mains. This project collected information on various available epoxy lining materials, their properties, historical performance, and longevity of epoxy liners. The key findings and recommendations of this project are provided in the following subsections.

FINDINGS

The key findings of this study are as follows:

- A survey of North American and United Kingdom water utilities conducted for this study indicates that only a few North American water utilities have experience using epoxy lining, whereas in the United Kingdom most of the water utilities used epoxy lining until 1998. Currently, polyurethane is used as the preferred lining material in the United Kingdom. Water quality concerns appear to be the prime driver of epoxy lining. All utilities responded that a significant water quality improvement was observed after lining. QA/QC procedures for construction of epoxy lining have been significantly improved during the last decade. All epoxies used for water main lining have NSF approvals, and AWWA is developing a standard for epoxy lining procedures. All utilities surveyed follow standard QA/QC procedures for epoxy lining installations. Water utilities have an expectation that epoxy linings will be effective for 30 to 50 years, and are satisfied with the performance of the liners.
- Epoxy liners, when constructed properly, can have a long life, which is estimated to be approximately 40 to 60 years based on this research.
- Controlled surface preparation and liner application are the most critical factors in ensuring an effective and long-lasting liner.
- Impedance, adhesion, and holiday density are important parameters for the longevity of a liner.
- Impedance is a good measure of the barrier properties of liner materials. High impedance values indicate good barrier properties. Impedance values are lower at higher temperatures for Type 1 liners.
- For Type 2 liners, low impedance values are observed initially at low temperatures and the values are increased with time (curing effect). In the long run, the impact of temperature on the impedance of a liner is less for Type 2 liners.
- Impedance can be measured accurately using EIS. Water utilities should consider measuring impedance to understand the barrier property of a liner.

- DFT is an important parameter for holidays and under-film corrosion. DFT is inversely related to holiday density and under-film corrosion.
- Historical epoxy liners with high impedance values and low holiday densities are found to be in excellent condition.
- Various application defects, including holidays and variations in DFT, have an impact on the longevity of a liner. These defects can be reduced with proper QC procedures.
- In laboratory panels, blistering was not observed on any sample.
- Corrosion of steel surfaces under the coating was related to barrier properties and holiday density.
- Well-constructed epoxy linings are generally very durable, and longevity may be more than 50 years.
- Poor construction of an epoxy liner has a severe impact on the longevity of the liner.
- Differences in epoxy lining material have minimum impacts on the longevity of a liner.

RECOMMENDATIONS

The following general recommendations are made as a result of this study:

- Epoxy lining for water mains has been found to be a good alternative to cement mortar lining and all North American water utilities should consider this technology as an alternative to cement mortar lining. All water utilities should consider conducting a cost-effective analysis of alternative linings before deciding on a lining material. Because of the thinner DFT and smoother lining surface, an epoxy lining will provide higher Hazen-Williams "C" values than a cement mortar lining. This may improve the hydraulic capacity of the lined water main, resulting in a reduction in energy consumption.
- Because proper QA/QC is a key in the success of good epoxy lining construction, it is imperative that water utilities strictly follow proper QA/QC procedures in order to obtain good longevity of the lined water mains.
- Water utilities should conduct CCTV surveys and measure Hazen-Williams "C" values of the water main before and after completion of the lining. Water utilities should keep proper records of the construction of the lining, and should take a coupon of the pipe sample after lining.
- For properly constructed epoxy linings, water utilities may expect a lining life of 40 to 60 years.
- Before selecting an epoxy lining material, water utilities should measure the impedance values of the liner. The higher the impedance of a liner, the higher will be the water barrier to the pipe wall. A liner with an impedance value of 10.5 to 11.0 will serve as good barrier.
- For a future study it is recommended that a detailed data collection effort be conducted relating holiday density, water quality, and customer complaints with the factor of safety. This would provide utilities with a better tool to estimate the factor of safety for estimation of the useful life of a liner.

APPENDIX A
WATER UTILITY QUESTIONNAIRE

APPENDIX A
WATER UTILITY QUESTIONNAIRE

1. Contact Information

Name of Organization ______________________________
Address ______________________________

Contact Name/Title ______________________________
Phone number ______________________________
Fax number ______________________________
E-mail address ______________________________

2. Epoxy lining history. Provide information in table below

Year of Epoxy Lining ______________________________

Project Name/Location ______________________________

Length of pipe lining ______________________________

Pipe diameter ______________________________

Pipe material ______________________________

Year pipe installed ______________________________

Epoxy manufacturer ______________________________

Epoxy product name ______________________________

Contractor name ______________________________

Cleaning method used ______________________________

Problems encountered during installation ______________________________

Target dry film thickness ______________________________

Mix ratio used ______________________________

Hardening time ______________________________

Time to full cure ______________________________

3. What was the motivation for your utility to provide epoxy lining?

Dirty water complaints ______________________________

4. In general, what has been the historical performance of the epoxy-lined pipe?

5. What has been your experience with pipe leaks?

5a. Have you experienced leaks in epoxy-lined pipes? (yes/no) ____________________

5b. If yes, what is the frequency of leaks? (leaks/year) ____________________

6. What has been your experience with pipe breaks?

6a. Have you experienced breaks in epoxy-lined pipes? (yes/no) ____________________

6b. If yes, what is the frequency of breaks in epoxy-lined pipes? (breaks/year) ____________

6c. What was the condition of the epoxy lining in the broken pipe? ____________________

7. What has been your experience in water quality changes?

7a. Have you experienced changes in water quality (either degradation or improvement) after epoxy lining? (yes/no) ____________________

7b. If yes, please describe ____________________

Better taste, fewer dirty water complaints ____________________

8. Have there been improvements in hydraulics?

8a. Have you experienced improvements in hydraulics (pressure, carrying capacity, C-factor)? (yes/no) ____________________

8b. If yes, please describe ____________________

9. What has been the condition of the epoxy lining?

9a. Have you inspected the condition of the epoxy lining after it was put back into service for a period of time? (yes/no) ____________________

9b. If yes, please describe ____________________

10. Have epoxy-lined pipe coupons been collected?

10a. Have you collected coupons of epoxy-lined pipe after it was lined? (yes/no) ____________

10b. If yes, describe the condition of the lining and the length of time that it had been in place.

____

10c. Were laboratory tests conducted on the lining from the coupon sample? (yes/no) ________

10d. If yes, please list the tests performed and the results:

Test ____

Result ____

Units ____

Test Standard Number/ID ____________________________

11. Are reports of the lining installation available? ____________________________

12. Are written installation procedures available? _______________________________

13. Is documentation kept of the condition of pipes before cleaning?

13a. Do you document the condition of the pipe before cleaning? (yes/no) _______________

13b. If yes, how do you document this? ____

14. What quality control procedures were used for cleaning and preparing the pipe surface? ____

15. Are written QC procedures for cleaning and surface preparation available?

____

16. What QC procedures were used for application of the epoxy liner?

____

17. Are written QC procedures for application of the epoxy liner available?

____

18. What QC procedures were used to evaluate the quality of the installed epoxy liner?

____

19. Are written QC procedures for post-installation evaluation available? ______________

20. What is your expectation of useful life of epoxy linings?

20a. Do you have an expectation of the useful life of the epoxy lining? (yes/no) ____________

20b. If so, what is your expectation for the useful life of the epoxy lining? (years) __________

21. Is the performance of epoxy lining satisfactory?

21a. Are you satisfied with the performance of the epoxy lining? (yes/no) _______________

21b. If no, provide explanation: ____

22. Are CCTV inspections performed?

22a. Do you perform CCTV inspection of the pipe before lining? (yes/no) ________________

22b. Do you perform CCTV inspection of the pipe after lining? (yes/no) ________________

23. What is the time duration between lining and putting the pipe back into service? (hours) ________________________________

APPENDIX B
EPOXY MANUFACTURERS' QUESTIONNAIRE

APPENDIX B
EPOXY MANUFACTURERS' QUESTIONNAIRE

1. Contact Information

Name of Organization ______________________________
Address ______________________________
Contact Name/Title ______________________________
Phone number ______________________________
E-mail address ______________________________

2. Name of epoxy product

____

3. Approval(s) for use of product in drinking water systems:

NSF Standard 61 certification? ______________________________

Certifying organization for NSF-61 (NSF or UL) ________________________

Regulation 25 (United Kingdom) ______________________________

Regulation 31 (United Kingdom) ______________________________

Other standard ______________________________

Country of jurisdiction ______________________________

4. Resin classification (bis-A or bis-F)

____

5. Solids content of resin (%)

____

6. Benzyl alcohol content (%)

____

7. Mix ratio (resin: hardener)

____

8. Recommended dry film application thickness (mm)

____

9. Recommended pipe temperature for application

____

10. Pre-heat temperature range for resin and hardener

____

11. Hardening time (hours)

____

12. Cure time (hours)

____

13. Time between application and return to service (hours)

____

14. Viscosity of resin at application temperature

____

15. Viscosity of hardener at application temperature

____

16. Material properties

Hardness ______________________________

Adhesion ______________________________

Impact resistance ______________________________

Flexibility ______________________________

Other (name) ______________________________

17. Long-term durability test

17a. Has there been any long-term durability testing performed on this product? (yes/no)

____

17b. If so, please describe:

____

18. Recommended surface preparation method

____

19. Recommended standard for surface preparation

20. Describe your recommended QA/QC procedures for lining inspection (e.g., video inspection, color, other testing procedures). Attach written procedures if available.

21. How many years has this product been in use in water pipe lining?

22. Current usage of epoxy products

22a. Is the epoxy product still in use in water pipe lining application? (yes/no)

22b. If not, name the product that replaced this product and complete a separate questionnaire for the replacement product.

APPENDIX C
DOCUMENTATION OF HISTORICAL PIPE SAMPLES

APPENDIX C
DOCUMENTATION OF HISTORICAL PIPE SAMPLES

SAMPLE #1 – YORKSHIRE WATER
(Applied in 1984 [4-inch cast iron])

Summary Comments:

- This lining was the second oldest of all of the samples received and was highly deteriorated.
- It had extensive blistering, with blisters up to ¾ inch in size.
- The lining contained continuous holidays during holiday testing.
- Many of the blisters had broken, and rust corrosion products were diffusing out of the blister.
- The film thickness of the lining averaged 19 mils, which was well below the required 40 mils.
- The lining also had extensive ringing, where the rings were about $\frac{1}{8}$ to $\frac{1}{4}$ inches wide.
- The blistered lining was removed so that the metal underlying the blisters could be examined. There was no significant mass loss associated with the blisters. Discoloration consisting of a film of reddish and yellowish corrosion products occurred under the blisters.
- The lining was brittle and showed shrinkage along some of the pipe wall.

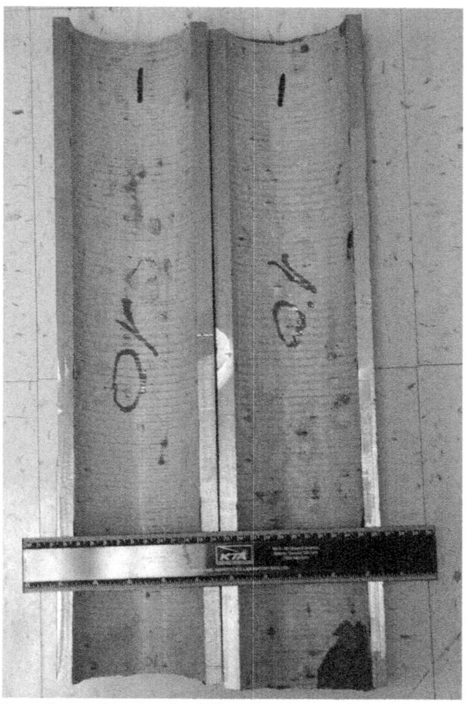

Figure C.1. Sample #1 as received, split in half. Split at 11:00 and 5:00 o'clock, with 12 o'clock in Section A (left). Note the highest density of blisters occurs at the 1 to 4 o'clock position.

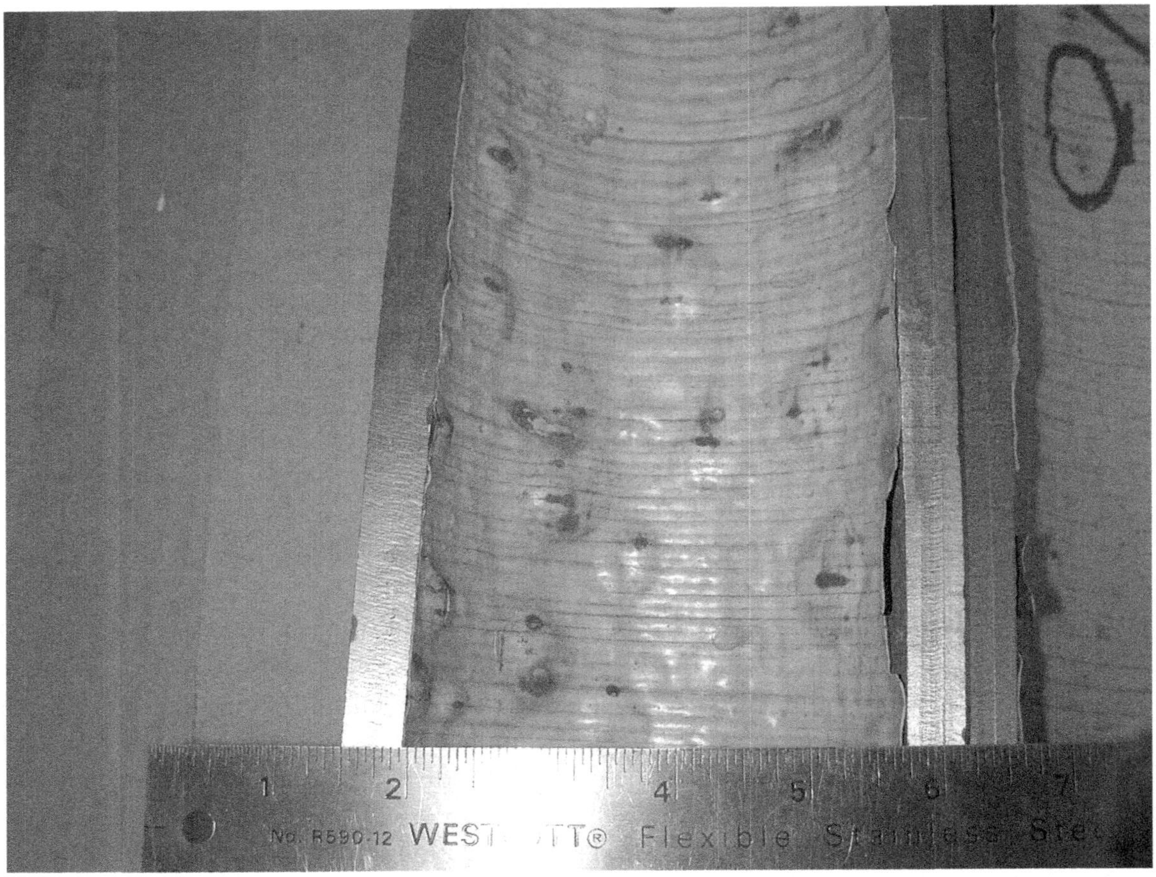

Figure C.2. Close up of split sample #1 showing blistering and rusting (including many blisters with broken domes). Note the ringing, severe deterioration, and shrinkage of the lining away from the pipe wall in one area.

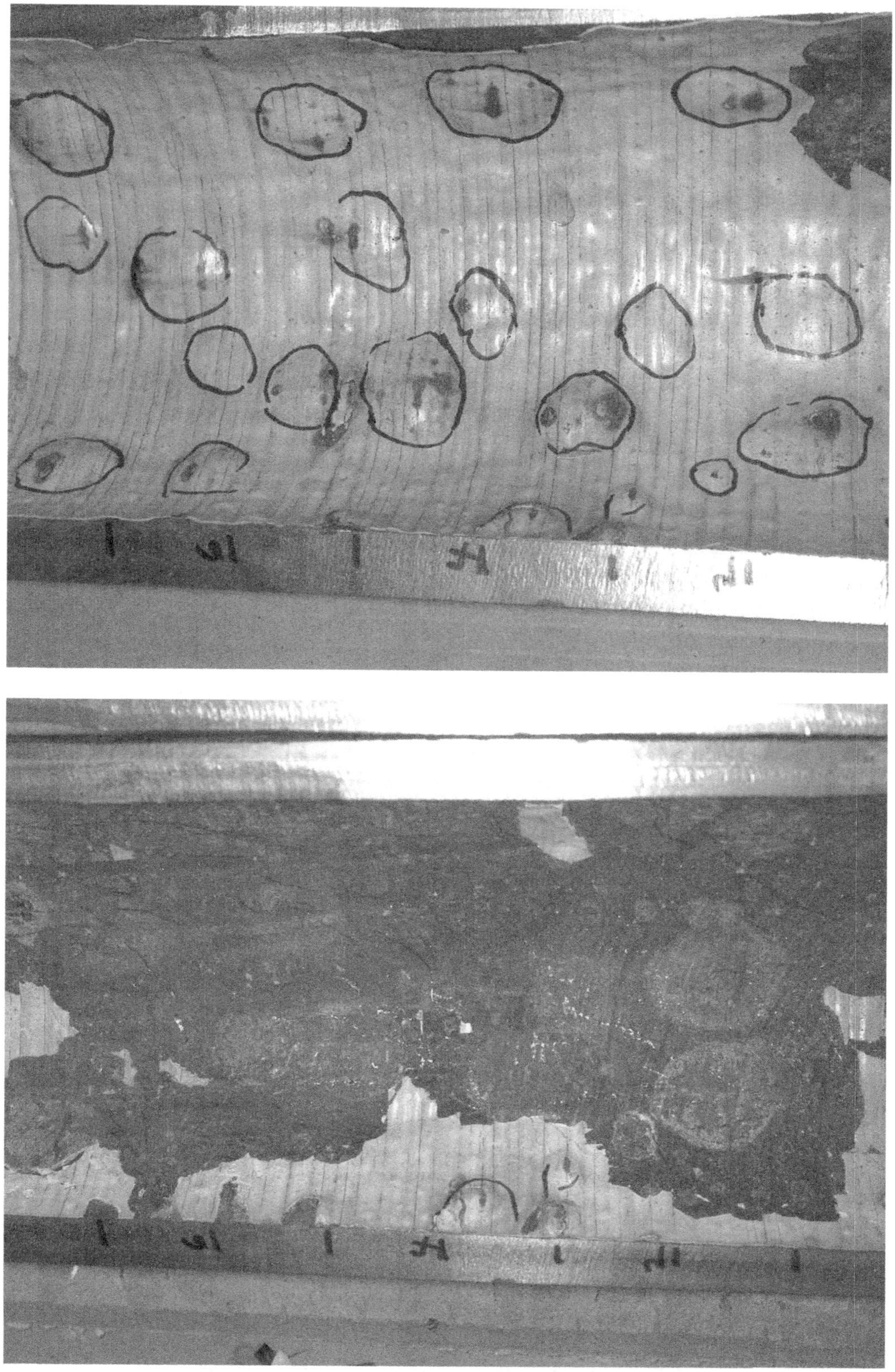

Figure C.3. Top: Close-up of blistered area of sample #1, with blisters circled.
Bottom: Same area after the coating was removed. Note the rust and corrosion marking the locations of many of the blisters. Mass loss was minor. Reference marks appear along the bottom edge of the pipe wall.

SAMPLE #3 – YORKSHIRE WATER
(Applied in 1991 [3-inch cast iron])

Summary Comments:

- This lining was 13 years old and in excellent condition with no blisters and one holiday.
- The film thickness of the lining averaged 32 mils, well below the required 40 mils.
- The finish was glossy with minor bleaching.
- The lining showed pooling, where the film thickness was 216 mils maximum at the measured area; the pooling presumably resulted in the overall reduced average thickness because the liquid lining flowed to the bottom of the pipe prior to gelling.
- The metal surface under the coating was the characteristic black observed in many of the samples. The surface contained ridges and markings likely produced during cleaning prior to lining application. The surface contained a small amount of black material, which could be scraped from the underlying solid material. The black material is presumably graphite. The scrape marks suggest the graphite was present at the time of lining.

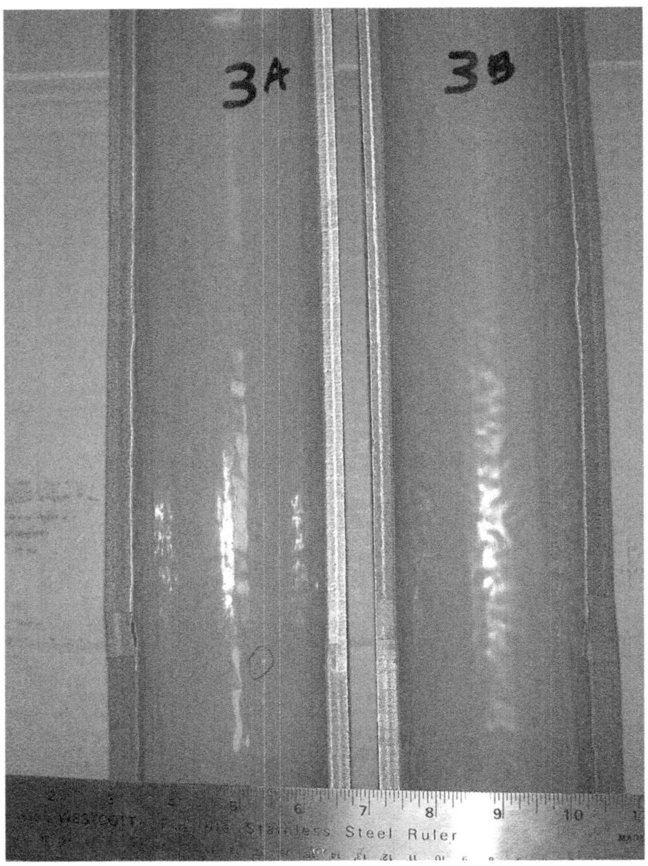

Figure C.4. Sample #3 as received, split in half, showing a holiday (circled). Split at 3 and 9 o'clock with 12 o'clock position in Section A (left). A pool is present at the top of Section B at 6 o'clock.

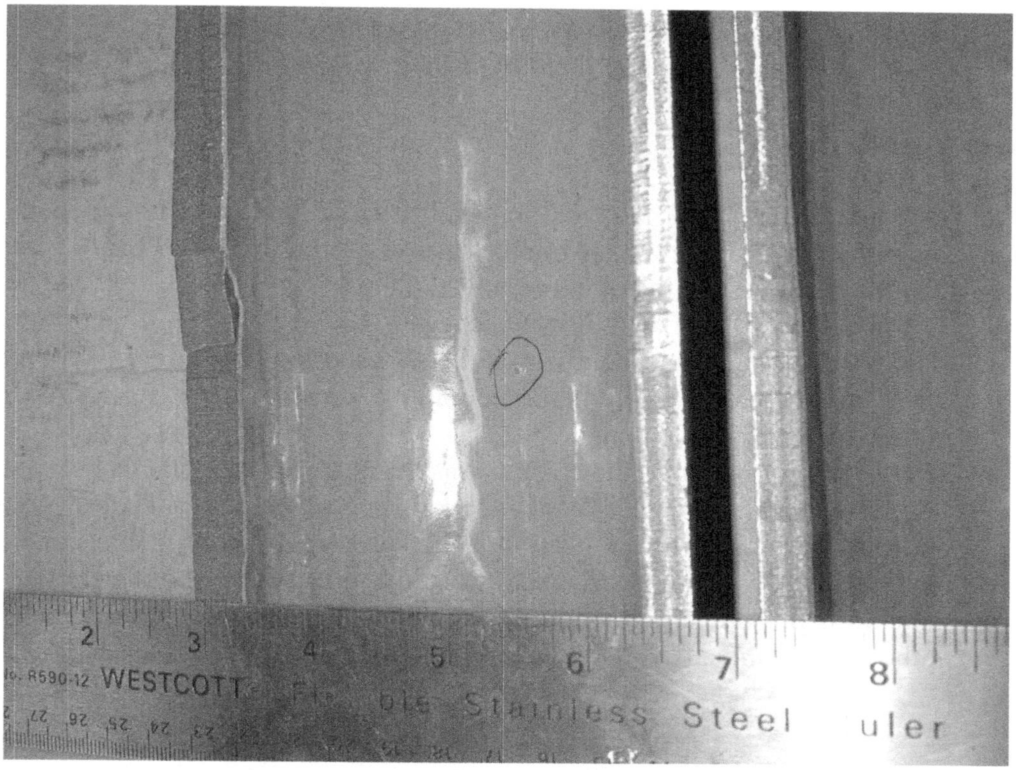

Figure C.5. Close up of split sample #3 showing the holiday. Note smooth surface.

SAMPLE #4 – YORKSHIRE WATER
(Applied in 1996 [4-inch spun iron])

Summary Comments:

- This lining was 8 years old and in excellent condition; it had no blisters, no ringing, and no holidays; there was no visible shrinkage from the pipe wall.
- The film thickness of the lining averaged 47 mils.
- The finish was glossy with little bleaching.
- Half of the lining was coated with a tightly adhering film of corrosion products from the 1 to 7 o'clock position, based on the crown stamp; inspection showed the underlying coating was sound.
- The metal surface under the coating was the characteristic black observed in many of the samples. The surface contained ridges and markings likely produced during cleaning prior to lining application. The surface contained a small amount of black material, which could be scraped from the underlying solid material. The black material is presumably graphite. The scrape marks suggest the graphite was present at the time of lining.

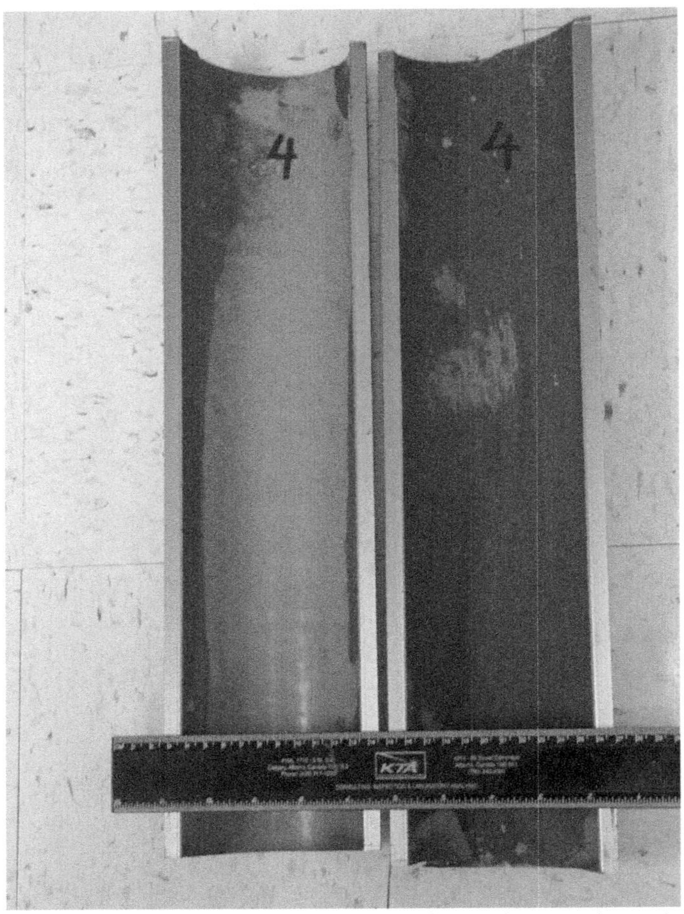

Figure C.6. Sample #4 as received, split in half, showing a heavy deposit of corrosion products (dark area) on half of the pipe. Split at 1 and 7 o'clock positions, with 12 o'clock position in Section A (left). Note rusted portion runs from 1 to 7 o'clock. The coating was fully intact under the corrosion product deposit. The deposit was conductive.

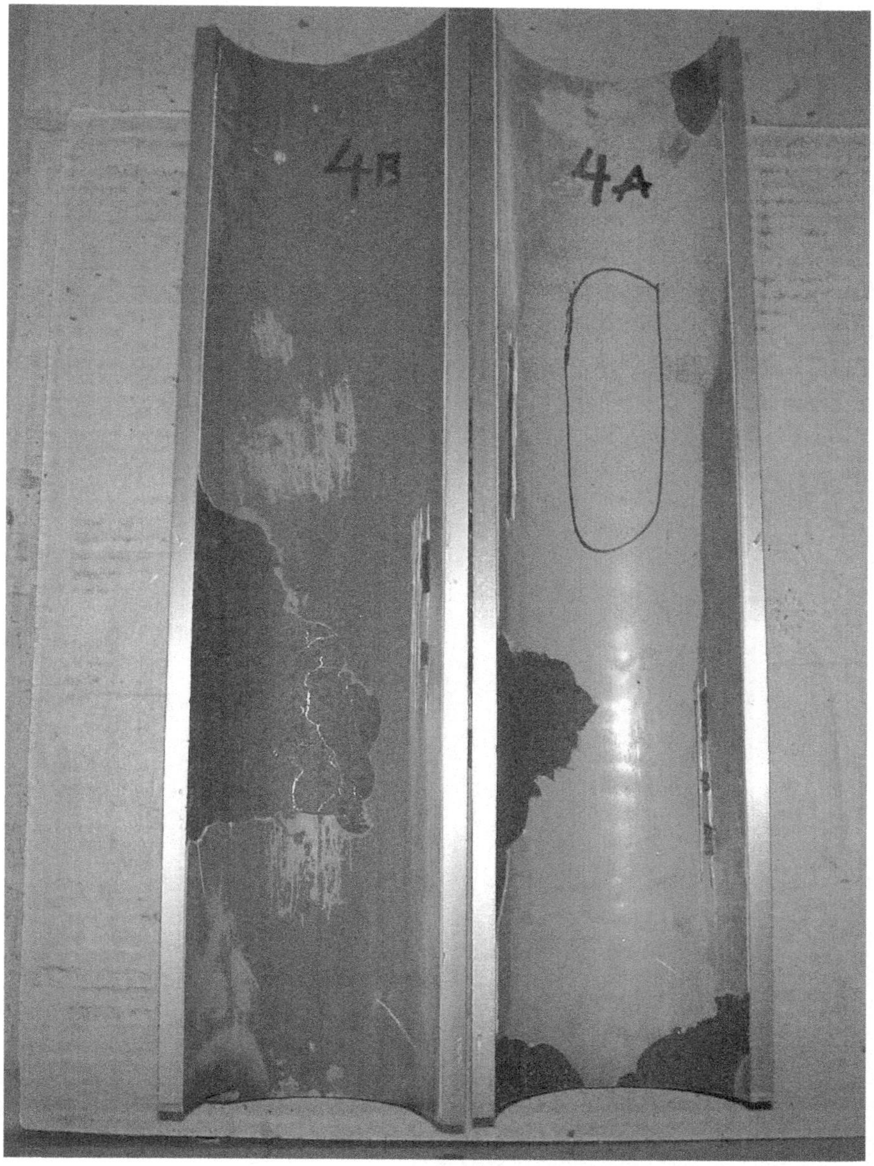

Figure C.7. Sample #4 pipe sections after some of the coating had been removed. Parallel scribe adhesion analyses can be observed. Circled area was used for EIS analysis.

SAMPLE #5 – YORKSHIRE WATER
(Applied in 1992 [3-inch cast iron])

Summary Comments:

- This lining was 12 years old and in generally very good condition.
- It had a few small isolated blisters, a somewhat irregular finish, and 7 holidays.
- The film thickness of the lining averaged 37 mils, slightly below the required 40 mils.
- The finish was glossy with minor bleaching.
- The lining had dark areas, which consisted of deposits of corrosion products or other material within the irregular surface finish.
- The lining was intact and without cracks at a location where a new service had been added.
- The metal surface under the coating had a thin film of red corrosion products over the characteristic black material observed in many of the samples.

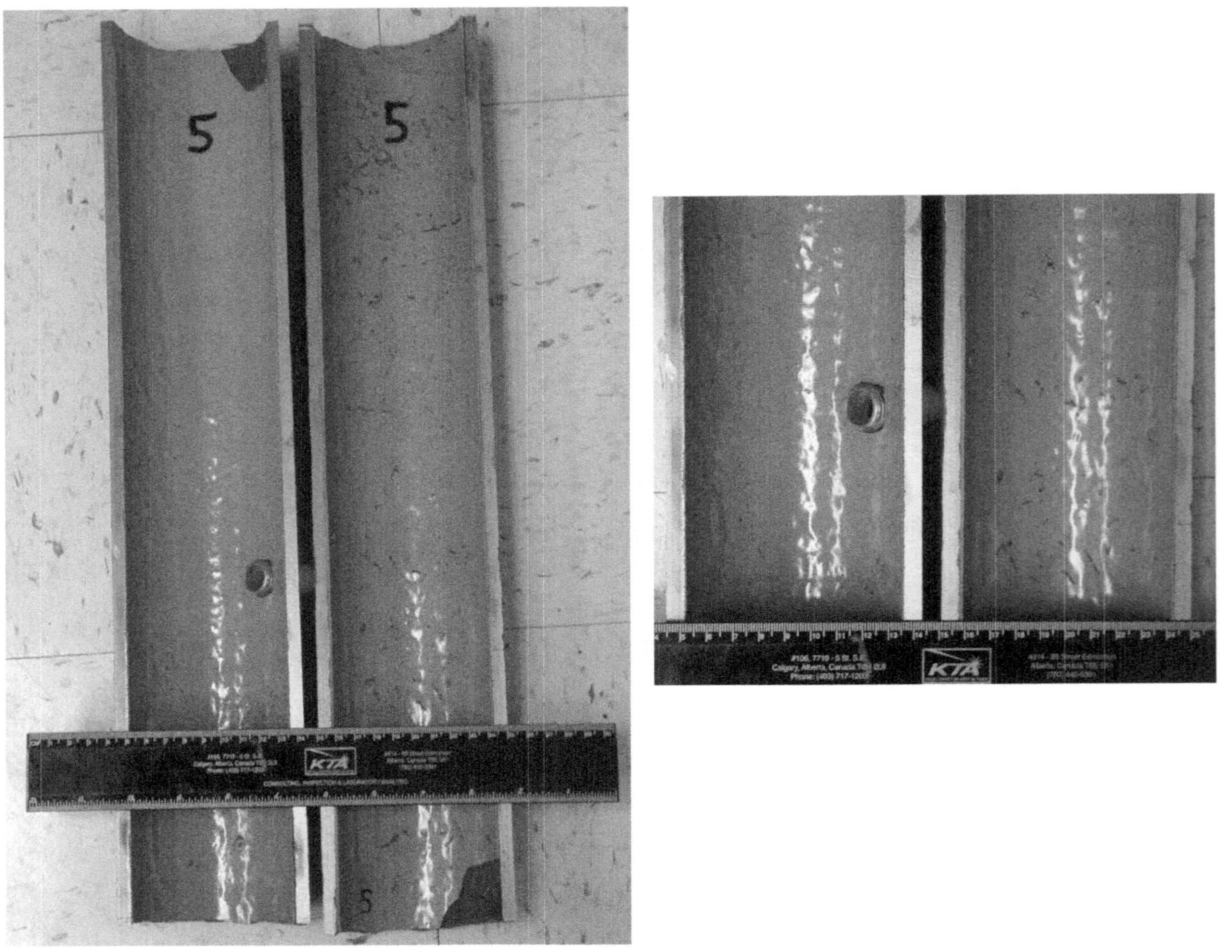

Figure C.8. Sample #5 as received, split in half, showing the irregular finish. Split at the 1 and 7 o'clock position, with 12 o'clock in Section A (right). Dark areas consist of a deposit of dark solid in the irregularities on the lining surface. Note the new service introduced after the lining was applied.

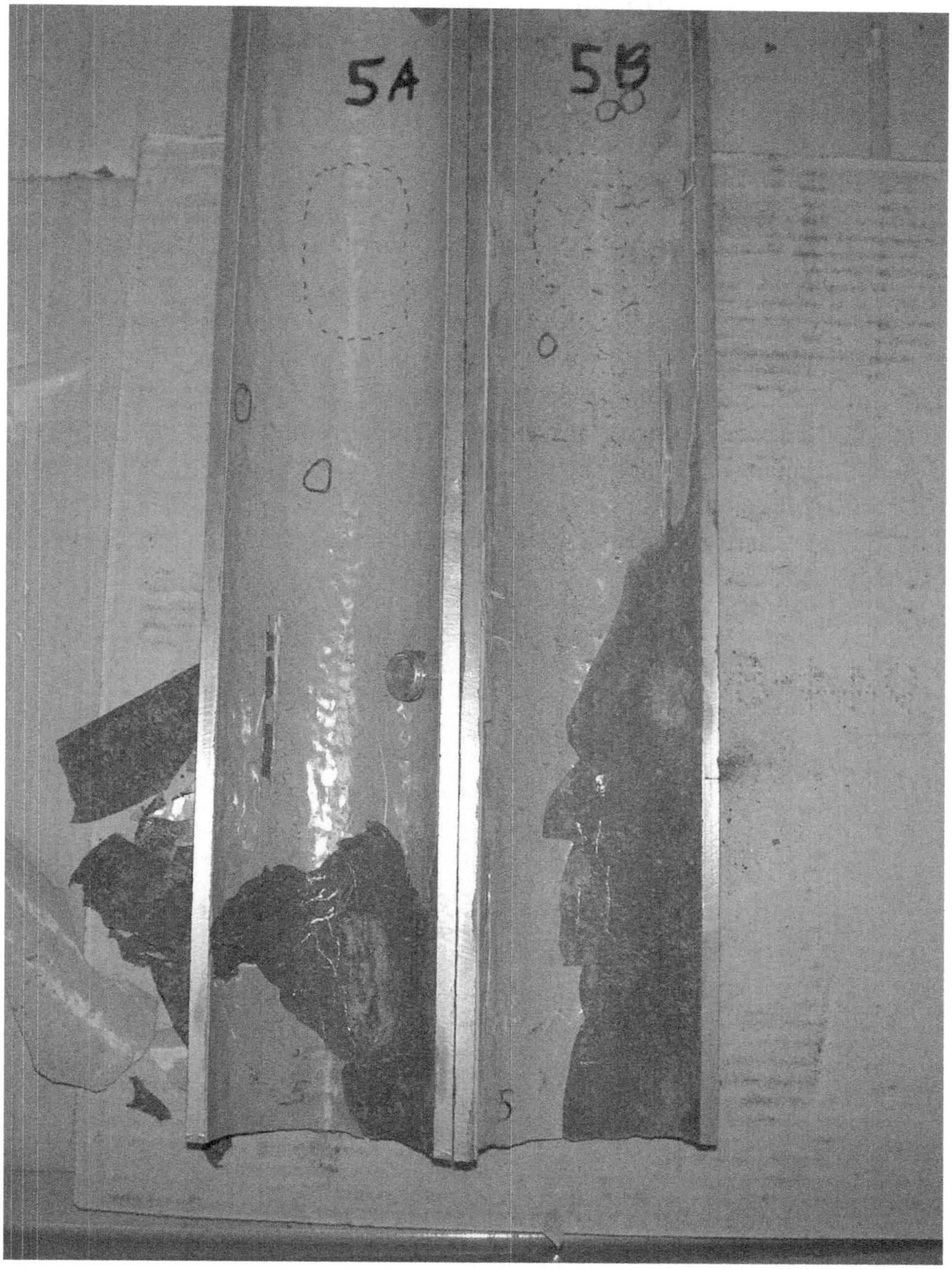

Figure C.9. Sample #5 pipe sections after analysis and after some of the lining was removed. Holidays are circled. Parallel scribe adhesion analyses can be observed. Large circled areas show location of EIS analysis.

SAMPLE #6 – YORKSHIRE WATER
(Applied in 2000 [4-inch cast iron])

Summary Comments:

- This lining was 4 years old, and in generally very good condition.
- It has 26 holidays but no blisters. The holidays ranged from large holes and volcanoes (1 mm openings) to microscopic, only detectable by holiday testing. Rust stains were not normally associated with the holidays.
- The film thickness of the lining averaged 47 mils, above the required 40 mils.
- The finish was glossy.
- The lining appeared to have pooling in one area, but investigation showed that the lining was only covering a flat residue of asphalt lining. The asphalt was presumably from a lining applied when the pipe was new. The residue did not appear to interfere with the application or function of the epoxy.
- The metal surface under the coating was the characteristic black, but with a thin film of reddish corrosion products.

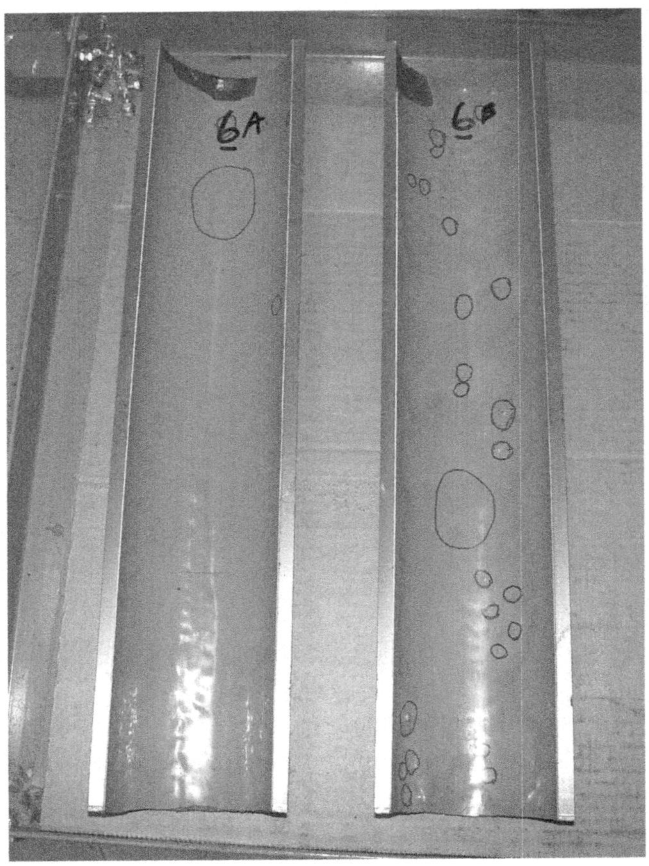

Figure C.10. Sample #6 as received, split in half, showing the somewhat irregular finish. Split at 2:30 and 8:30 positions, with Section B having the 12 o'clock position. Note that the holidays are concentrated around the 12:30 position and pooling over the asphalt residues is in the middle of Section A. Holidays are marked with circles. Large circles show holiday-free areas marked for EIS measurements.

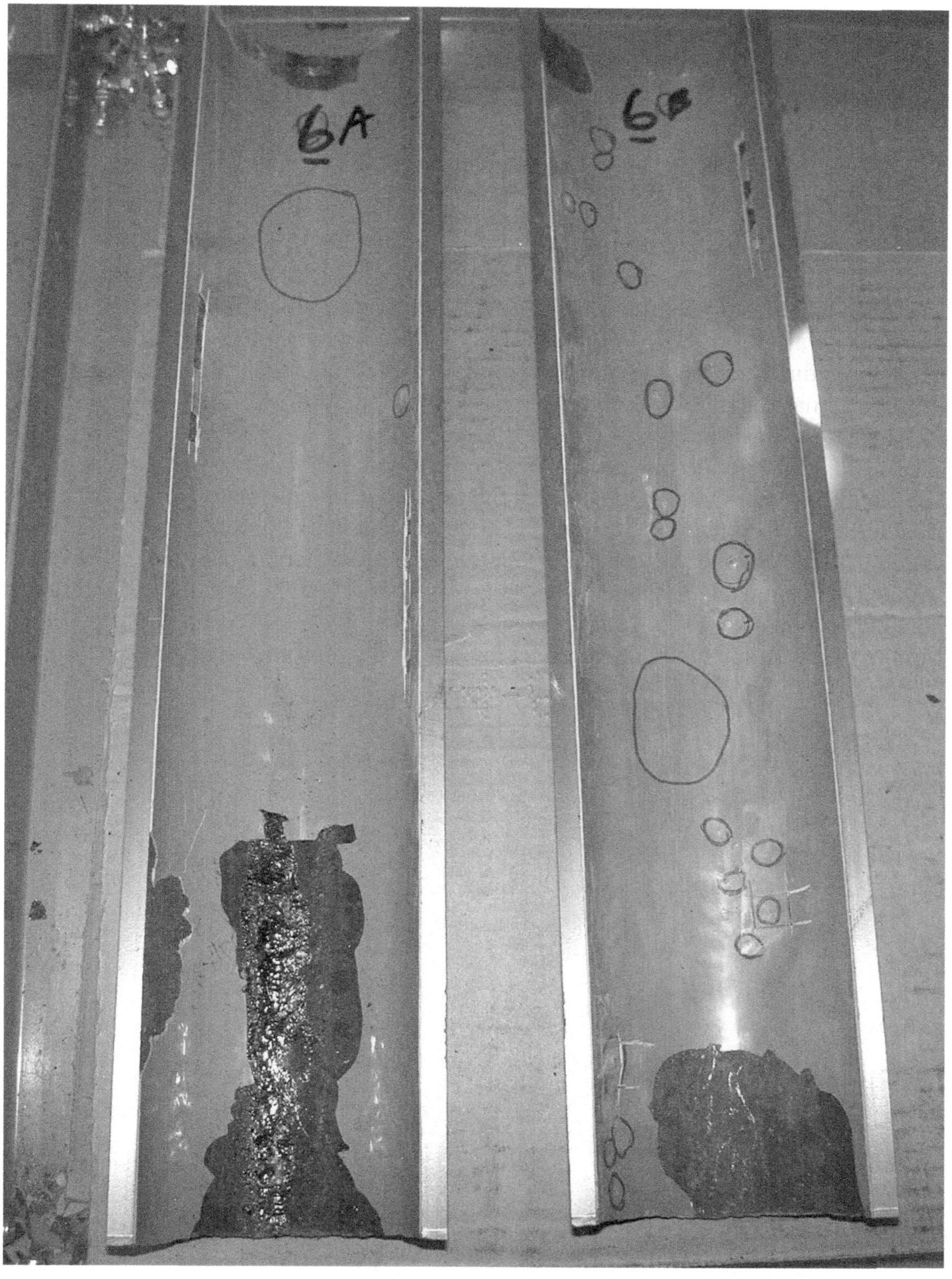

Figure C.11. Sample #6 pipe sections after analysis and removal of some of the coating. Holidays are circled. Parallel scribe adhesion analyses can be observed. Large circled area was used for EIS analysis.

Note the residue of asphalt lining on the pipe surface in the picture on the left. The asphalt produced some of the irregularity (apparent pooling) seen on the epoxy surface. Note the absence of metal loss. Some rust is present. The bright streaks are from the prying action of the utility knife to remove the lining.

SAMPLE #8 – YORKSHIRE WATER
(Applied in 1995 [3-inch cast iron])

Summary Comments:

- This lining was 9 years old and in good condition in spite of an irregular finish and an obvious longitudinal defect.
- Detailed examination, however, showed that it had 8 holidays, mostly associated with the longitudinal defect. There were no blisters. Pooling was observed in part of the sample.
- The film thickness of the lining averaged 44 mils, close to the required 40 mils.
- The finish was glossy.
- The longitudinal defect, upon investigation, consisted of a sharp ridge on the surface of the original casting. The epoxy lining was unable to bridge the gap produced by the ridge, resulting in holidays.
- The metal surface under the coating was the characteristic black observed in many of the samples. The surface contained ridges and markings likely produced during cleaning prior to lining application. The surface contained a small amount of black material, which could be scraped from the underlying solid material. The black material is presumably graphite. The scrape marks suggest the graphite was present at the time of lining.
- Minor graphitization could be observed in parts of the pipe wall cross section.

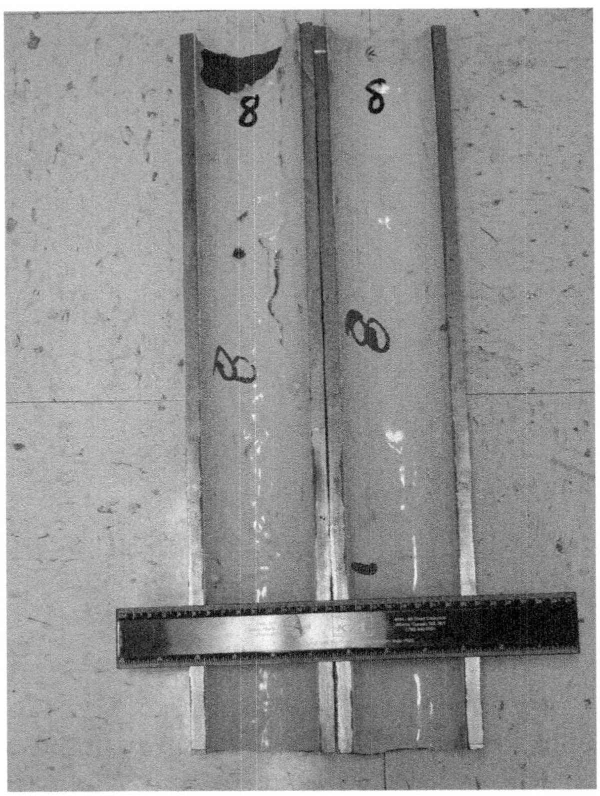

Figure C.12. Sample #8 as received, split in half, showing the very irregular, but glossy finish. Split at 3:30 and 9:30 positions. Section A (left) includes the 12 o'clock position, which includes a longitudinal defect consisting of a gap in the coating and rough surface. Pooling can be observed in the middle of Section B (right).

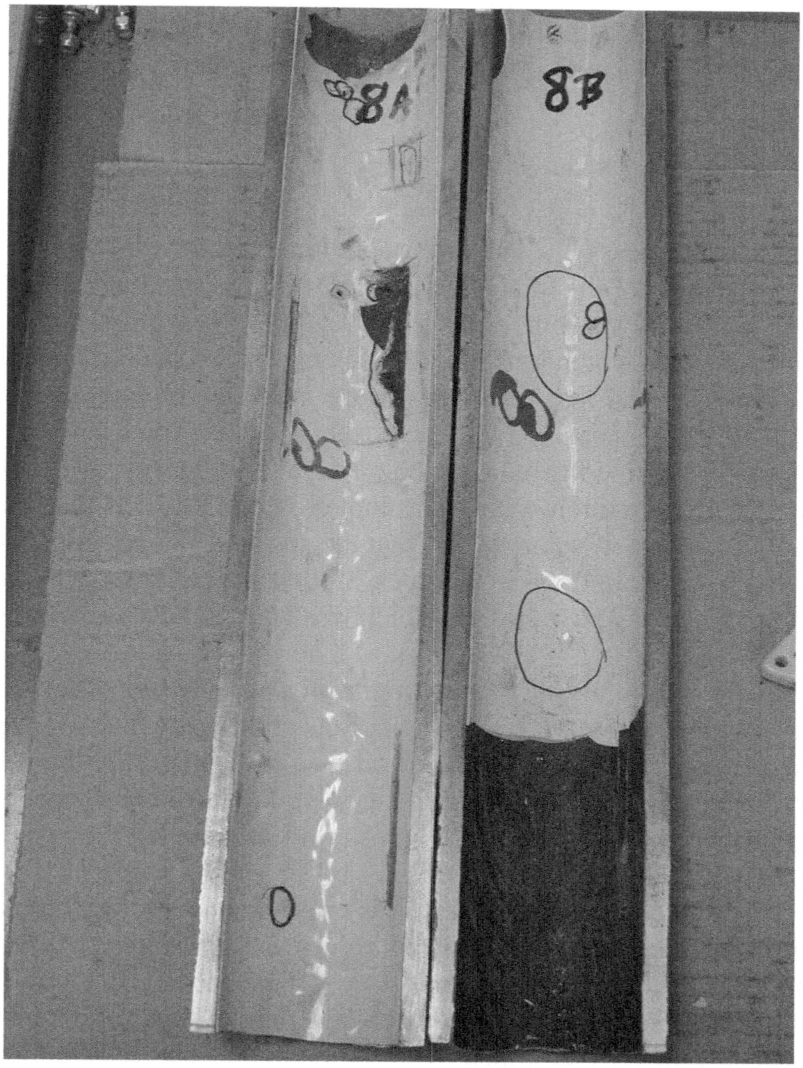

Figure C.13. Sample #8 pipe sections after analysis and removal of some of the lining. Holidays are circled. Areas with large circles were used for EIS analysis. Parallel scribe adhesion analyses can be observed.

SAMPLE #9 – YORKSHIRE WATER
(Applied in 1995 [4-inch cast iron])

Summary Comments:

- This lining was 9 years old and in fair condition with no blisters.
- The film thickness of the lining averaged 32 mils, below the required 40 mils.
- The finish was glossy but pocked, with minor bleaching; the surface finish had numerous speckles which looked like black embedded material in the surface.
- Investigation showed that the lining had numerous holidays, including an area that consisted of continuous holidays (i.e., the holiday detector sparked continuously). The area with continuous holidays also had a high density of speckles. Contrary to expectation, investigation with a stereomicroscope showed that the coating film in this area was not visibly porous and that the speckles did not consist of black material distributed throughout the film thickness. Therefore, the porous nature is attributed to microscopic, non-visible porosity, which was undoubtedly aggravated by the low film thickness.
- In this sample, a rare instance of rusting from a holiday was observed.
- Some pooling was observed, along with runs and sags, which probably contributed to the overall low film build-up. Based on the crown marking, the pooling occurred at the 3 o'clock position, suggesting the crown may have been miss-marked. The area with continuous holidays occurred across from the pooled area, suggesting it was, in reality, at the 12 o'clock position.
- The metal surface under the coating was the characteristic black observed in many of the samples. The surface contained ridges and markings likely produced during cleaning prior to lining application. The surface contained a small amount of black material, which could be scraped from the underlying solid material. The black material is presumably graphite. The scrape marks suggest the graphite was present at the time of lining.
- The lining was brittle and cracked during dry out, an effect likely aggravated by the low film thickness.

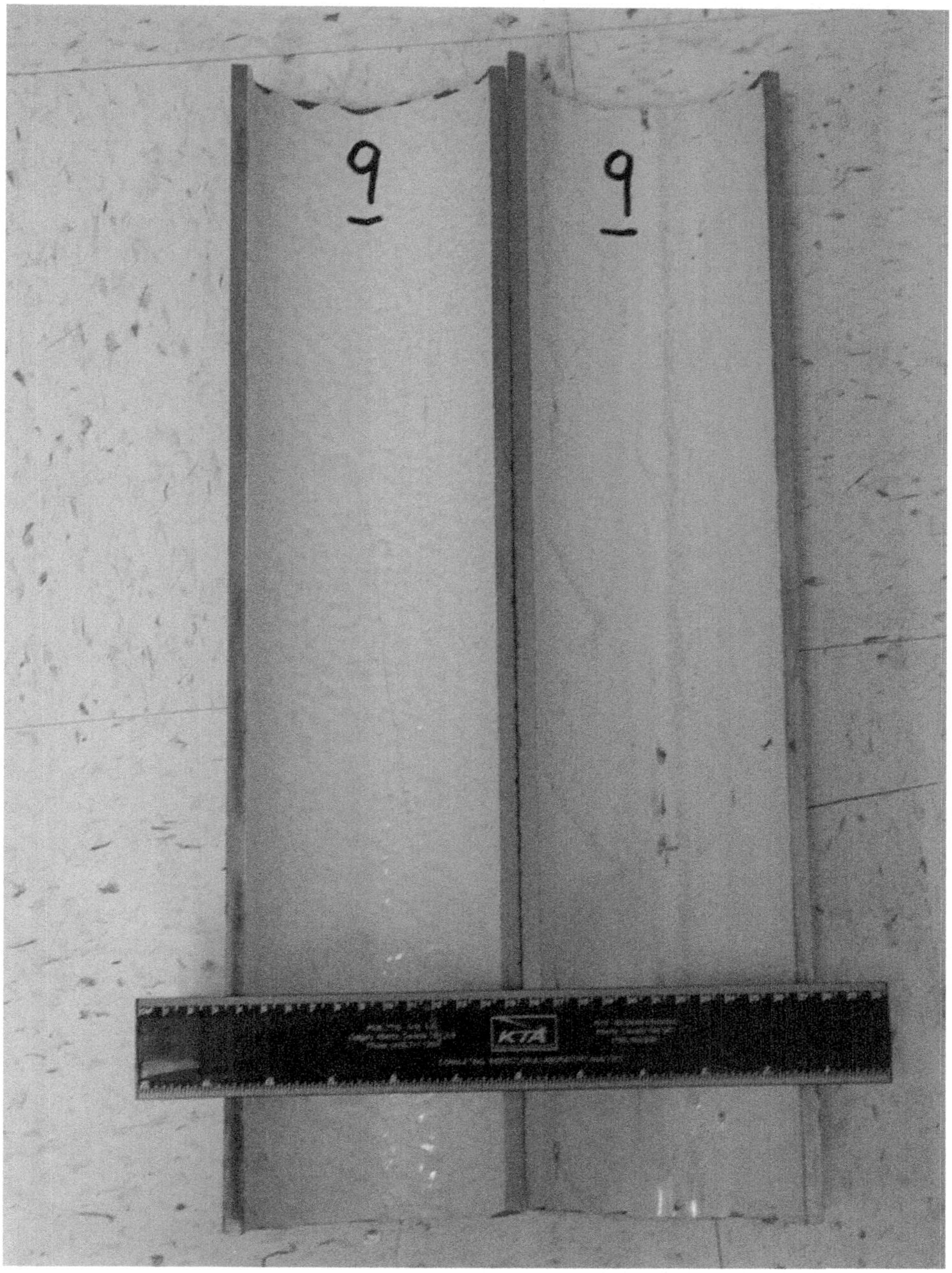

Figure C.14. Sample #9 as received, split in half, showing a rough finish. The cracks occurred after the lining dried out. Split at the 12 and 6 o'clock positions, based on the position stamped at the crown (12 o'clock, which is inconsistent with pooling, runs, and sags). More likely orientation is 3 and 9 o'clock, with Section B (left) with high porosity including the 12 o'clock position. Note the pooling in the middle of Section A (right).

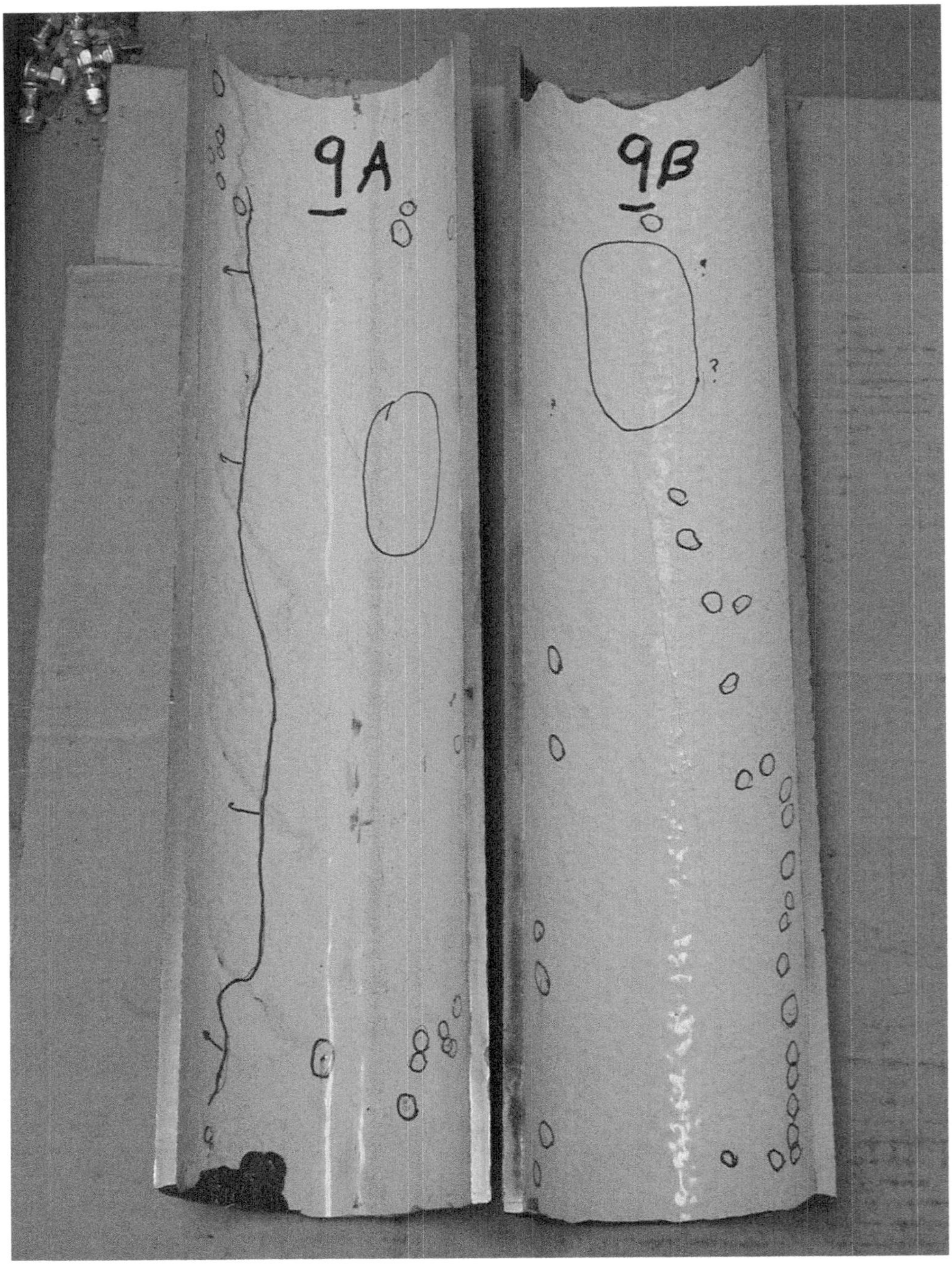

Figure C.15. Sample #9 pipe sections showing numerous holidays. The zone marked along the left side of the section on the left was porous and contained a high density of continuous holidays. Areas with large circles were used for EIS analysis.

SAMPLE #10 – YORKSHIRE WATER
(Applied in 1995 [4-inch cast iron])

Summary Comments:

- This lining was 9 years old and in fair condition with no blisters.
- The film thickness of the lining averaged 62 mils. Due to severe ringing, however, the actual film thickness ranged from 20 mils to 134 mils at the thin and thick areas of the rings.
- The surface had lost most of its gloss, and bleaching was highly variable, depending on film thickness in the rings.
- The thin areas of the rings showed continuous holidays in holiday testing.
- The metal surface under the lining was the characteristic black observed in many of the samples, with a thin film of red iron oxide in some areas. There was no relationship between the pattern of ringing and the corrosion of the underlying base metal.
- The lining showed shrinkage away from the pipe walls, associated with dry out.

Figure C.16. Sample #10 as received, split in half, showing severe ringing.

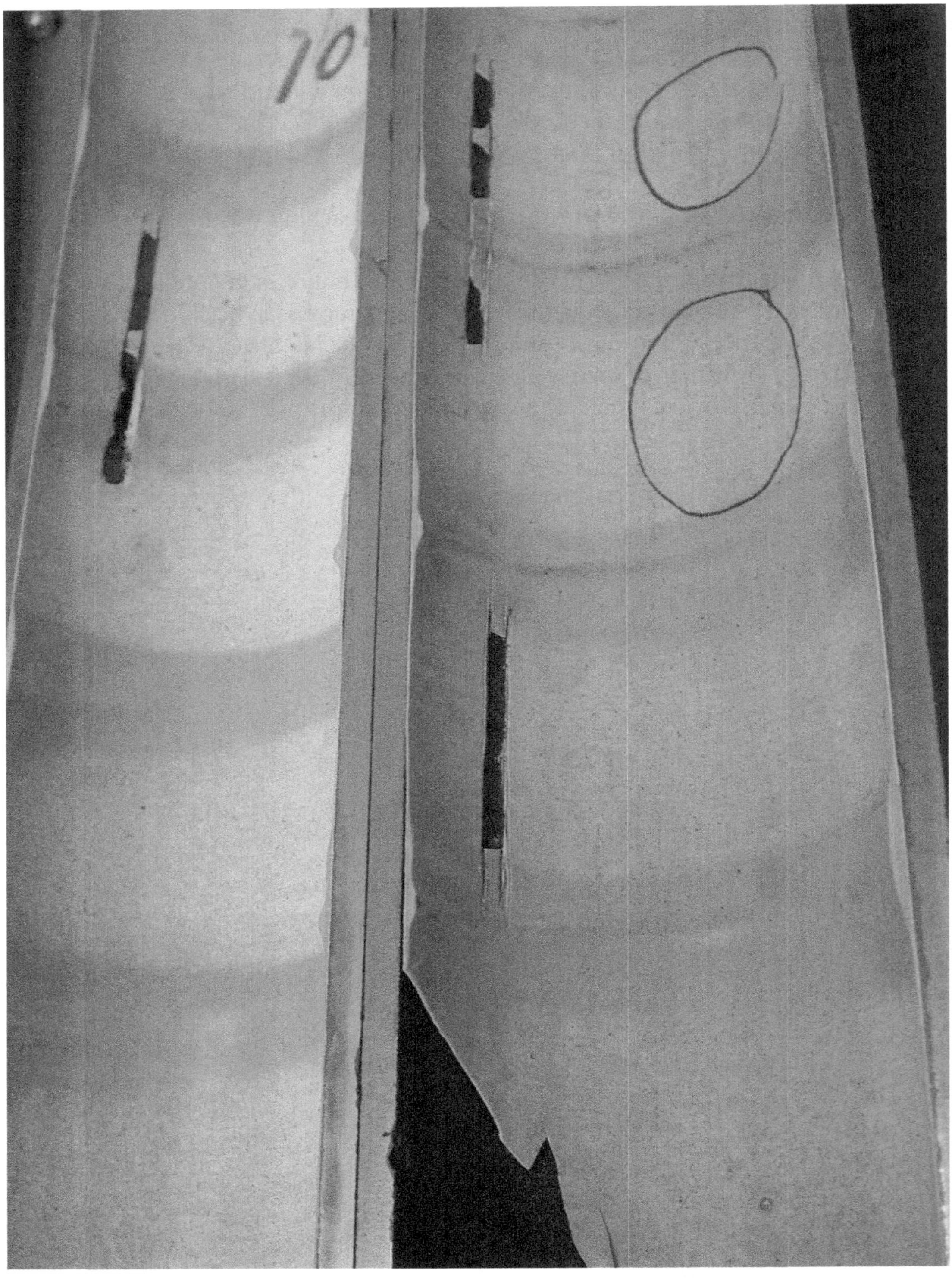

Figure C.17. Sample #10 pipe sections after parallel scribe analysis. Areas with large circles were used for EIS analysis. Parallel scribe adhesion analyses can be observed.

SAMPLE #11 – YORKSHIRE WATER
(Applied in 1995 [3-inch cast iron])

Summary Comments:

- This lining was 9 years old and had no blisters and no holidays.
- The film thickness of the lining averaged 56 mils, well above the required 40 mils.
- The surface had lost some of its gloss, and showed some bleaching.
- The metal surface under the lining was the characteristic black observed in many of the samples. The metal surface was in excellent condition with little, if any, mass loss.
- The lining showed shrinkage along some of the pipe wall after dry out.

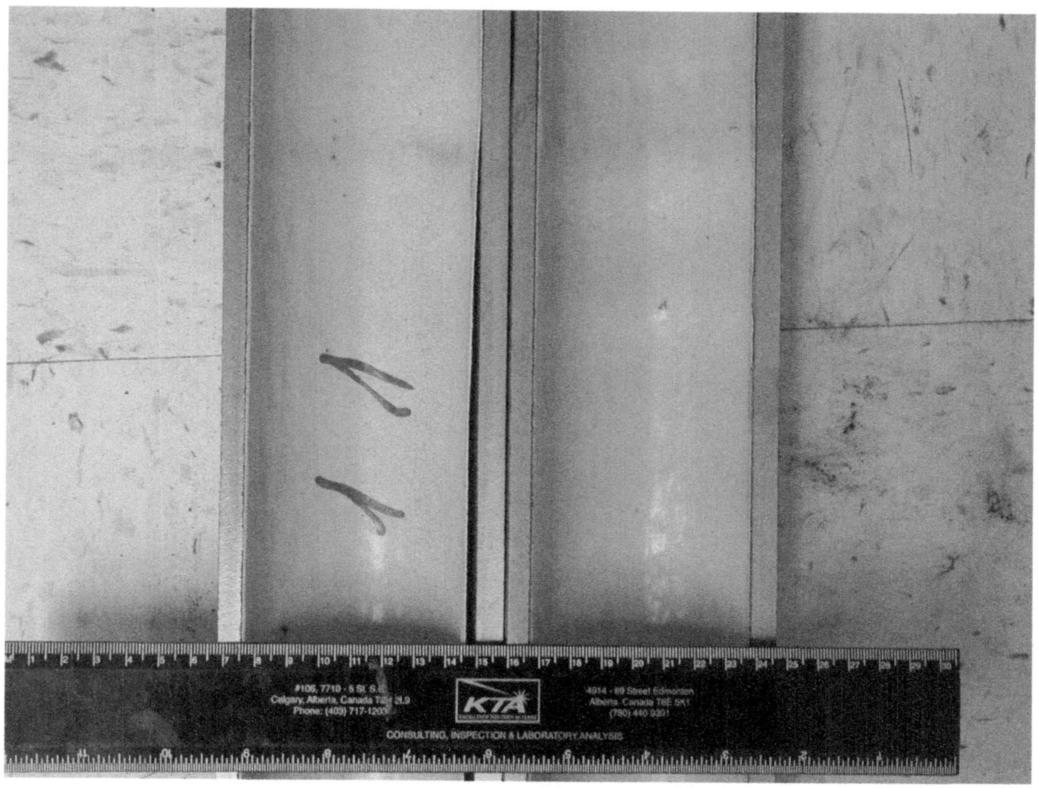

Figure C.18. Close-up of pipe sample #11 surface, showing good quality finish.

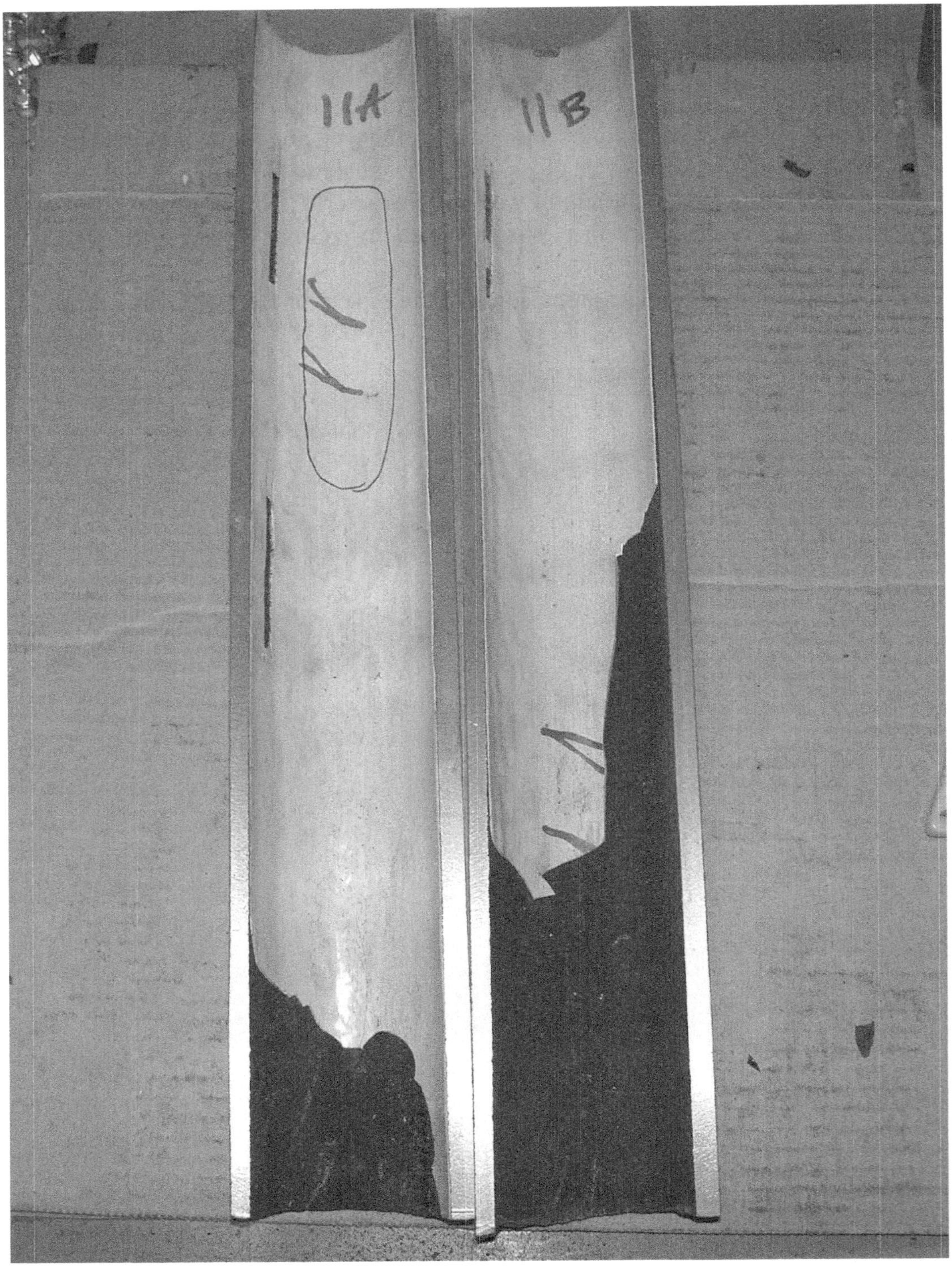

Figure C.19. Sample #11 pipe sections after analysis and removal of some of the coating. No holidays were detected. Areas with large circles were used for EIS analysis. Parallel scribe adhesion analyses can be observed.

SAMPLE #12 – YORKSHIRE WATER
(Applied in 1979 [4-inch cast iron])

Summary Comments:

- This lining was 25 years old, the oldest of all of the samples examined.
- The lining was in a highly deteriorated state with large, rust-filled tubercles.
- When holiday tested, it had continuous holidays.
- The film thickness of the lining averaged 38 mils, slightly below the required 40 mils.
- The lining finish was matte and bleached. The application was smooth, however, and without visible ringing.
- The metal surface under the lining and adjacent tubercles was a mixture of black with reddish and yellowish corrosion products. There was no visible or measurable mass loss under the lining or tubercles.
- The non-tuberculated lining did not have any blisters.
- The lining was gummy in nature, and when the lining was cut with a Dremel tool for parallel scribe adhesion analysis, the lining melted rather than dusting like the other epoxies; this suggests a possible application problem (e.g., in mixing or curing).
- The lining is an early version of Geopox (Geopox GX012), which was superseded by Geopox GX014.

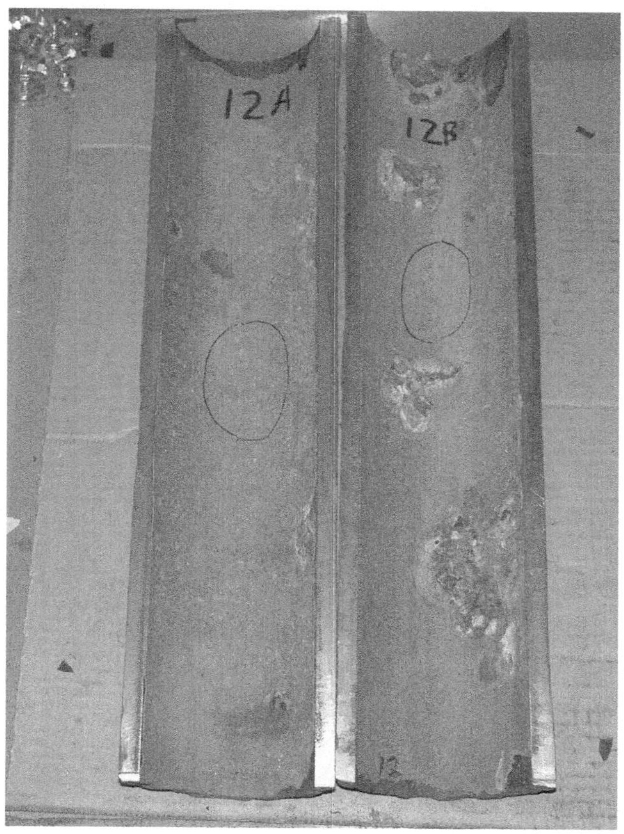

Figure C.20. Sample #12 as received, split in half, showing the severe deterioration and large tubercles. Split at the 1 and 7 o'clock positions. Tubercles on right half (B) are at the 3 to 6 o'clock position.

SAMPLE #101 – DENVER WATER
(Applied in 2001 [6-inch cast iron])

Summary Comments:

- This lining was 3 years old and in excellent condition with no blisters and three holidays.
- The film thickness of the lining averaged 54 mils, well above the required 40 mils.
- The finish was glossy with slight bleaching.
- The lining was very smooth on one half of the pipe section, and rippled on the other side. Possibly the rippled side was at the 12 o'clock position, and the rippling resulted from a drip-like sag of the lining prior to gelling.
- The lining had been applied over cement mortar. The lining had little adhesion to the mortar after dry out, and each half of the lining could readily be lifted away from the mortar surface as a single piece. The liner appeared to have experienced some shrinkage during dry out, which would have increased the tendency to disbond from the mortar.
- The mortar under the lining was in excellent condition.
- A thin layer of black material was observed between the mortar and the epoxy lining, possibly affecting the adhesion between the liner and mortar.

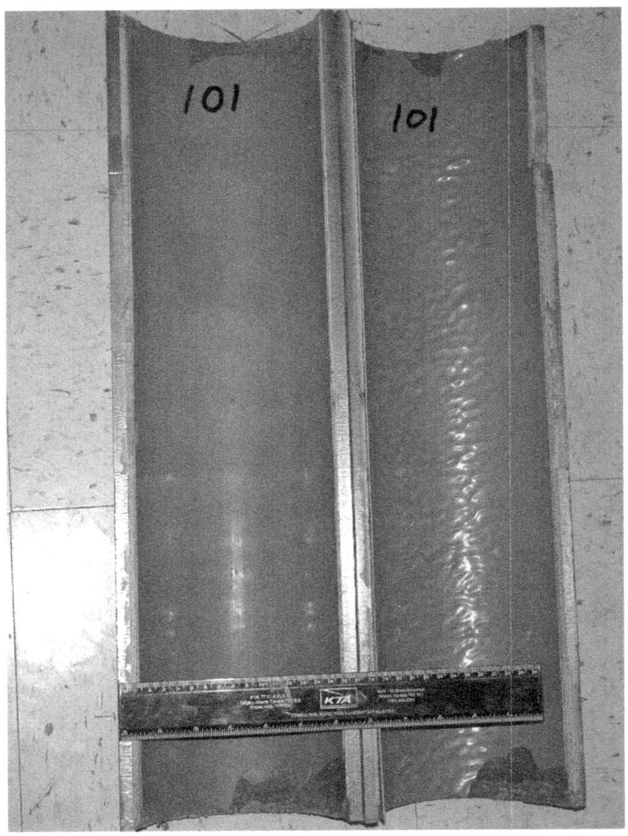

Figure C.21. Sample #101 as received, split in half, showing the rippled finish on half of the sample. Sample split at 3 and 9 o'clock positions, with A (half on the right with drip/ripple marks), including the 12 o'clock position. Note that this sample consisted of the epoxy lining applied over a cement mortar lining.

Figure C.22. Sample #101 pictures showing that the entire epoxy liner can be lifted in one piece from the pre-existing cement mortar lining.

SAMPLE #102 – DENVER WATER
(Applied in 1998 [6-inch cast iron])

Summary Comments:

- This lining was 6 years old and in excellent condition with no blisters and one holiday.
- The film thickness of the lining averaged 63 mils, well above the required 40 mils.
- The finish was glossy with minor bleaching.
- The lining showed ringing, and a few runs, sags, and dribbles.
- The metal surface under the lining was the characteristic black observed in many of the samples. The surface contained ridges and markings likely produced during cleaning prior to lining application. The surface contained a small amount of black material, which could be scraped from the underlying solid material. The black material is presumably graphite. The scrape marks suggest that the graphite was present at the time of lining.
- This lining did not shrink during dry out, or pull away from the pipe wall.

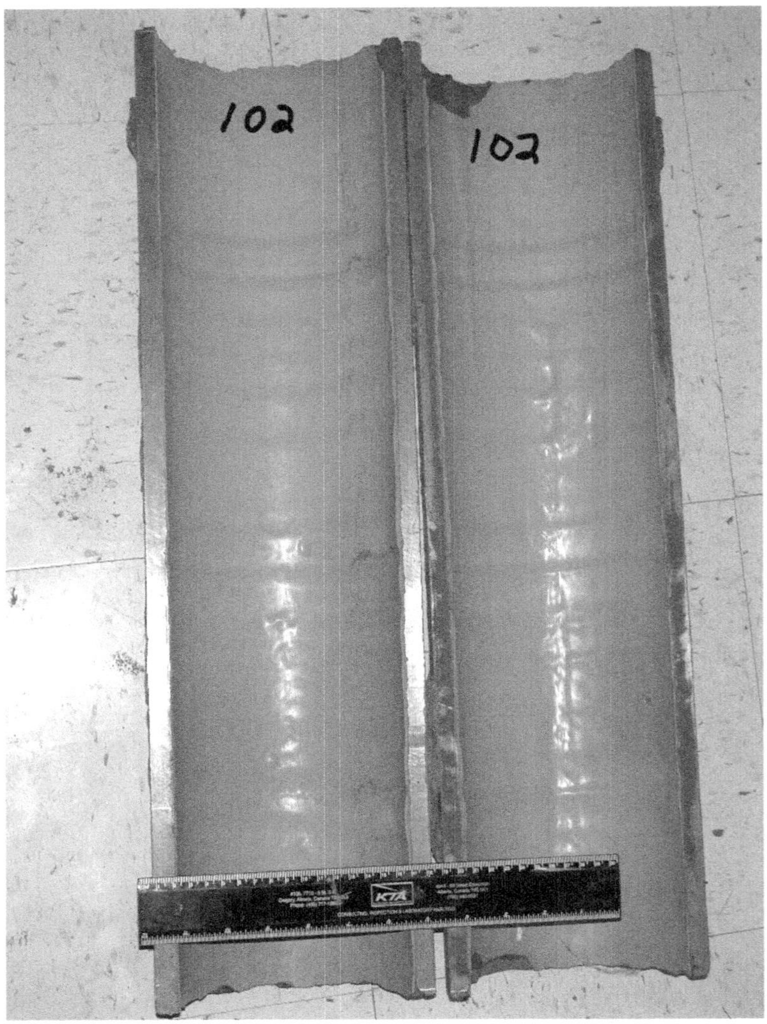

Figure C.23. Sample #102 as received, split in half, showing the ringing and irregular finish. Spit at the 6 and 12 o'clock positions.

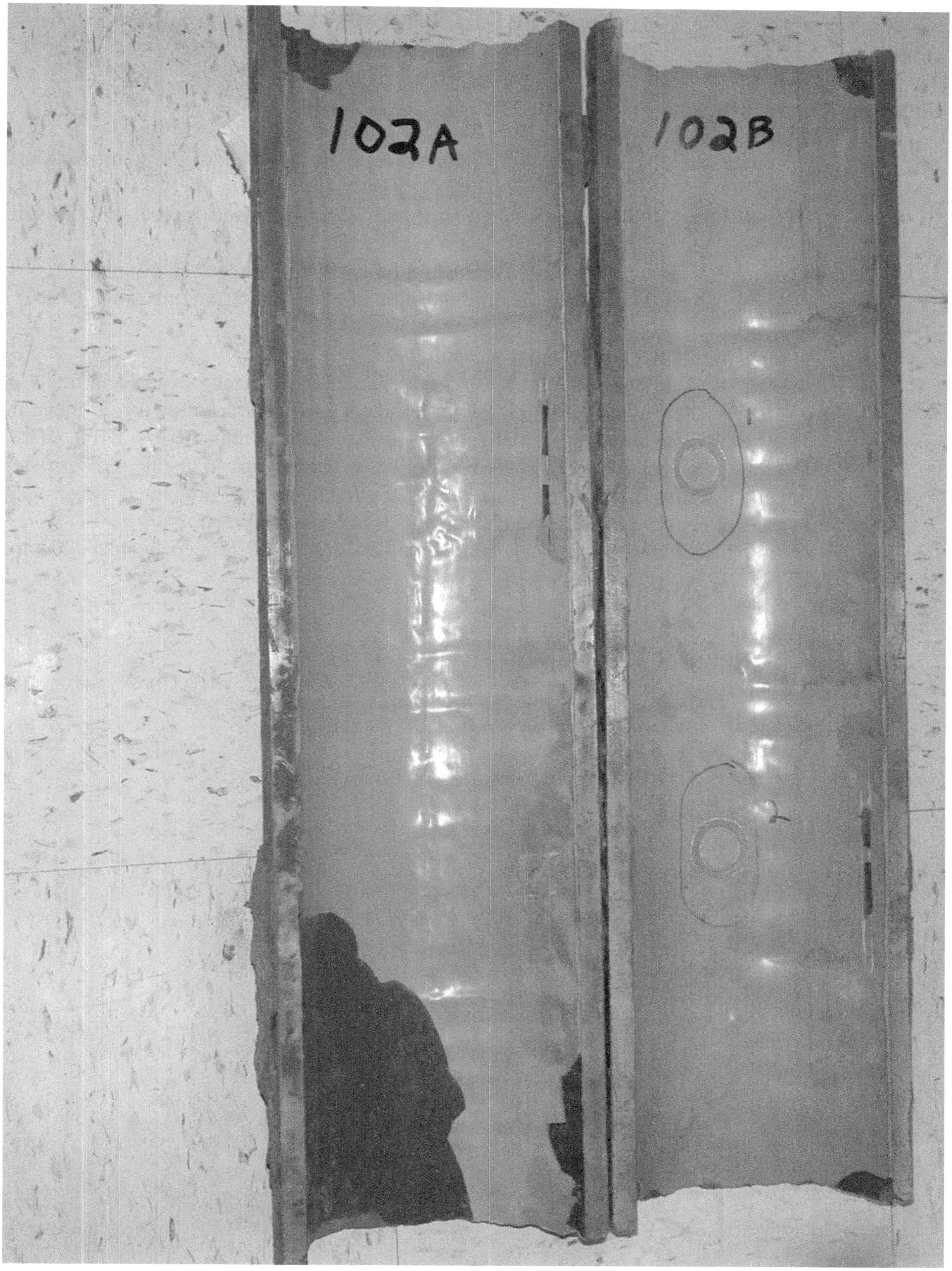

Figure C.24. Sample #102 pipe sections after analysis and removal of some of the lining. Areas with large circles were used for EIS analysis.

SAMPLE #103 – DENVER WATER
(Applied in 1999 [6-inch cast iron])

Summary Comments:

- This lining was 5 years old and visually in relatively good condition with some rust staining.
- Detailed analysis, however, showed that it had 77 holidays and several large blisters (1 to 1.5 inches in size).
- The film thickness of the lining averaged 53 mils, above the required 40 mils.
- The lining showed considerable ringing, with film thickness ranging from 19 to 87 mils.
- The finish was glossy with minor bleaching.
- The metal surface under the lining was the characteristic black observed in many of the samples. Removal of the lining over holidays and blisters, following by cleaning of the surface (to remove graphitized cast iron), showed that there was no visible metal loss associated with the blisters or holidays. There were no visible red corrosion products associated with the blisters. The metal surface contained ridges and markings likely produced during cleaning prior to lining application.
- The lining showed no shrinkage and remained tightly positioned against the pipe wall.

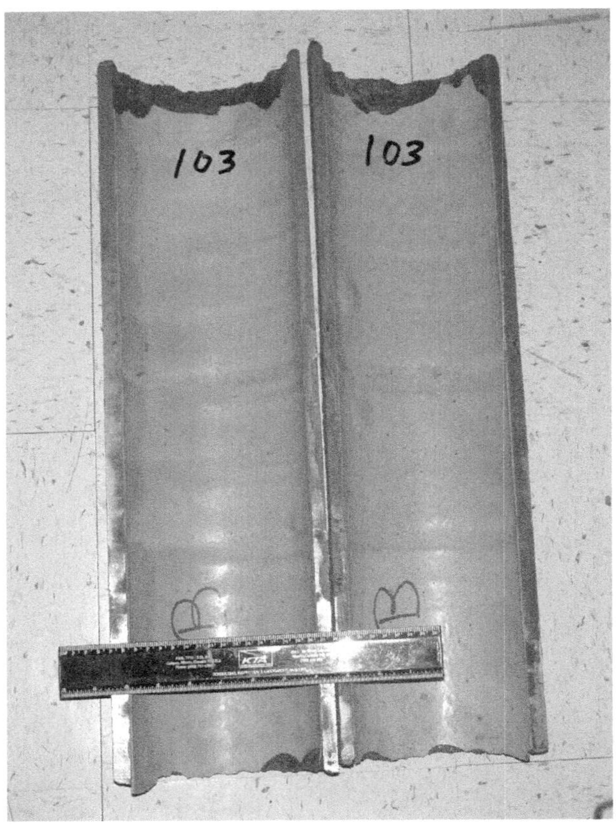

Figure C.25. Sample #103 as received, split in half, showing significant ringing. Split at 1 and 7 o'clock positions. Section B (on the left) includes the 12 o'clock position. Note the two circular areas (left side of right section toward the top), which look like blisters.

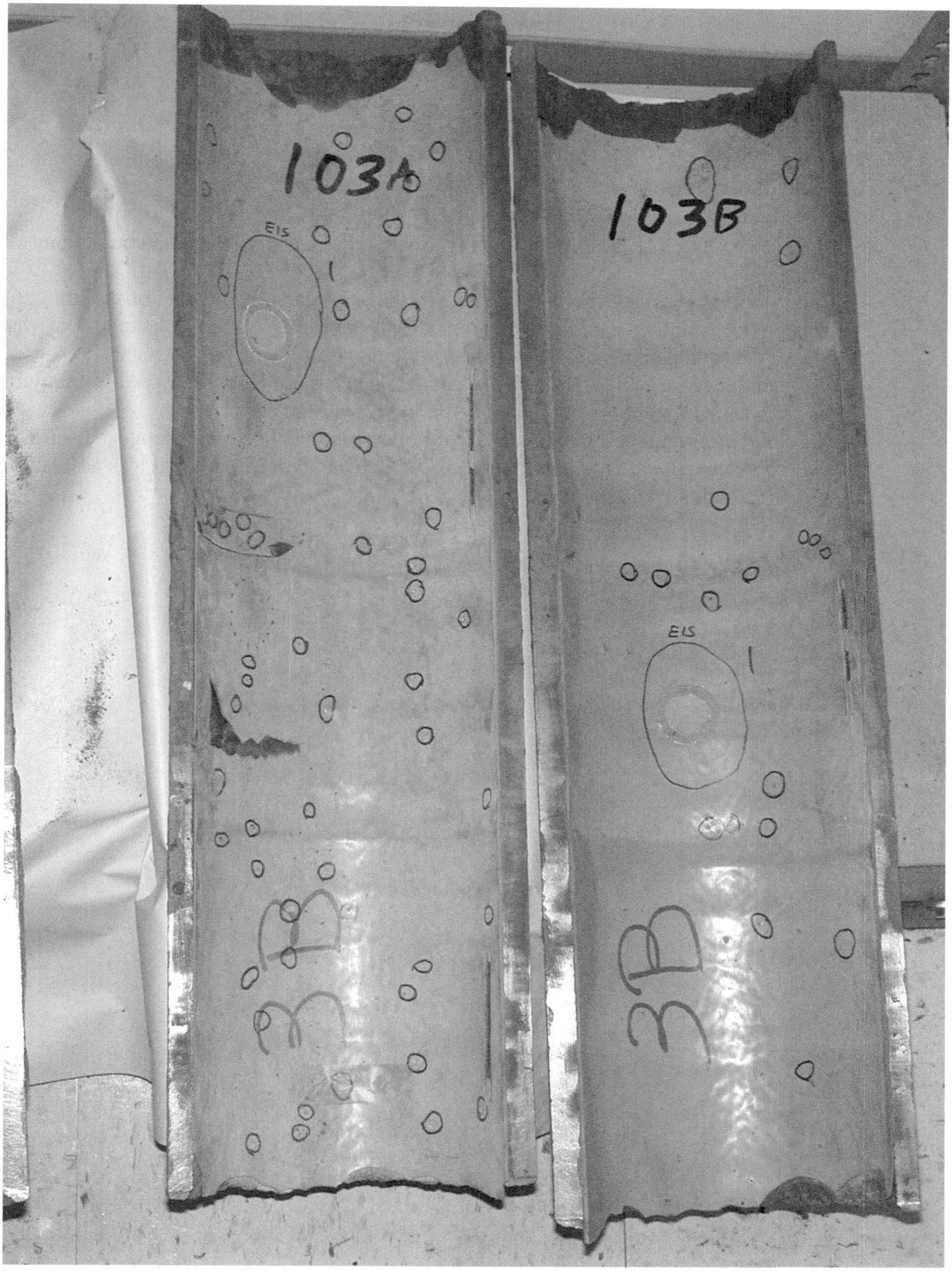

Figure C.26. Sample #103 pipe sections after marking holidays (circled). Areas with large circles were used for EIS analysis.

SAMPLE #104 – CHESTER WATER
(Applied in 1993 [6-inch cast iron])

Summary Comments:

- This lining was 11 years old and visually in fair condition. It had no blisters but 17 holidays.
- The film thickness of the lining averaged 47 mils, above the required 40 mils.
- The surface had lost its glossy finish and had experienced a considerable amount of surface bleaching.
- Upon initial inspection, the lining appeared to have a large, elongated blister about 6-inches long. Further examination showed that the lining was not blistered, but was covering a rough, elongated deposit of casting material on the pipe surface.
- The metal surface under the lining was the characteristic black observed in many of the samples.
- The lining was removed over a cluster of holidays so that the underlying steel surface could be cleaned and examined in detail. There was no visible metal loss associated with the holidays.
- The lining did not show any shrinkage away from the pipe wall.

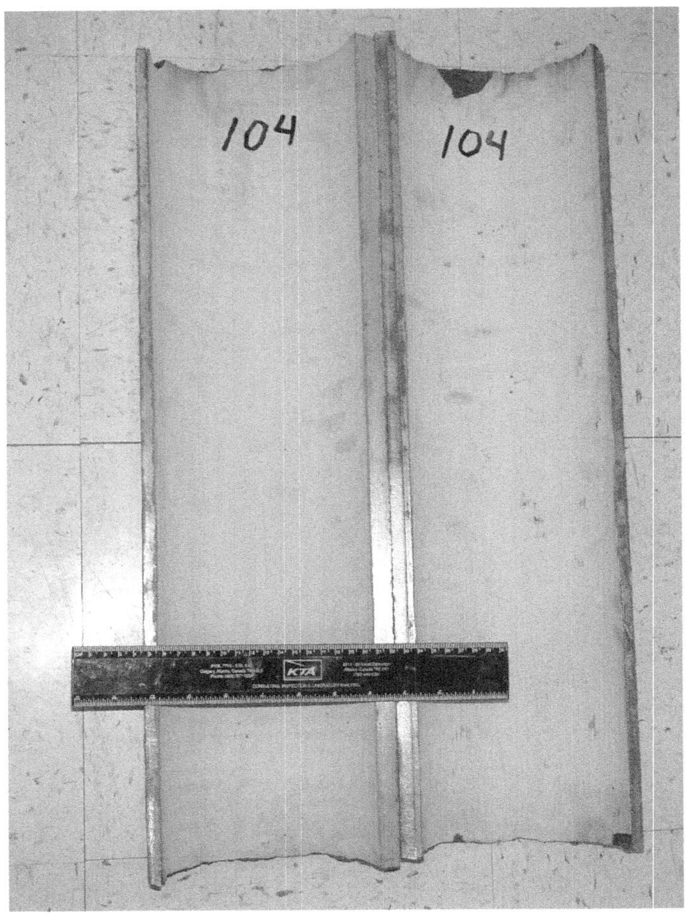

Figure C.27. Sample #104 as received, split in half, showing the lining surface. Orientation unknown.

Figure C.28. Sample #104 holidays marked on the epoxy lining.

SAMPLE #105 – HALIFAX REGIONAL WATER
(Applied in 2000 or 2002 [6-inch cast iron])

Summary Comments:

- This lining was between 2 and 4 years old and was in generally excellent condition with no blisters and three holidays.
- The film thickness of the lining averaged 52 mils, above the required 40 mils.
- The finish was glossy; a thin film of adherent corrosion products was present in a 1-inch side strip along the length of the pipe; the corrosion products presumably originated from upstream of the lining.
- The metal surface under the lining was the characteristic black observed in many of the samples with a light film of red corrosion products. There was no observable metal loss associated with the strip of corrosion products or the holidays in the lining.
- Varying degrees of penetration of graphitization from both the inside and outside were observed on the cut edges of the pipe.

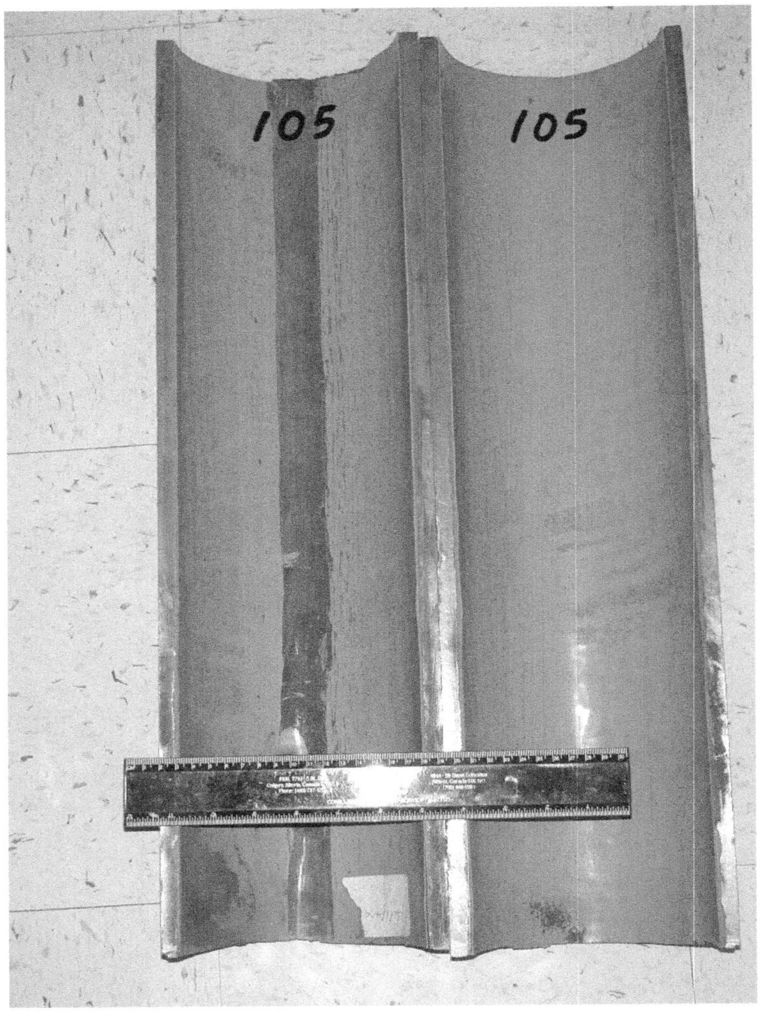

Figure C.29. Sample #105 as received, split in half, showing a linear deposit of tightly adhering corrosion products. Orientation unknown.

Figure C.30. Sample #105 pipe sections after analysis and removal of some of the coating. Holidays are circled. Parallel scribe adhesion analyses can be observed.

SAMPLE #106 – HALIFAX REGIONAL WATER
(Applied in 2003 [12-inch cast iron])

Summary Comments:

- This lining was 1 year old and initially appeared to be in very poor condition. The lining had many rust stains on its surface and appeared to have many blisters. Detailed examination showed that the lining had 25 holidays, only some of which had rust stains associated with them. The rust stains occurred primarily on intact lining and were presumably deposited from corrosion products generated upstream in an unlined portion of the pipe. Many of what looked like blisters were actually runs and dribbles of epoxy that occurred during application.
- The lining had a cluster of 5 blisters, which were 1 to 1.5 inches in size.
- The film thickness of the lining averaged 49 mils, close to the minimum 40-mil requirement.
- The finish was glossy with minor bleaching.
- The metal surface under the lining was the characteristic black with a thin film of red corrosion products, which varied in amount from place to place. The pattern of red corrosion products was not related to the location of holidays or blisters. There was no significant metal loss associated with the blisters or holidays.
- The impedance of intact areas of the lining had log Z values greater than 10, indicating that the holiday-free areas of the lining were in excellent condition with excellent barrier properties.

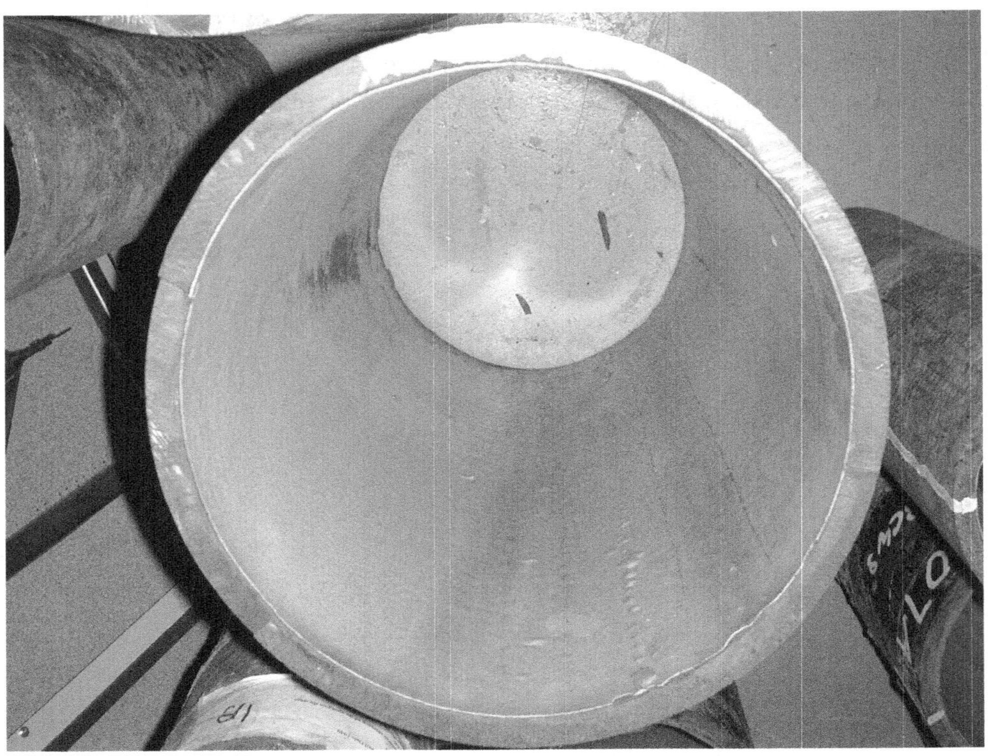

Figure C.31. Sample #106 as received. Dribble marks in pipe toward bottom of the picture are at about the 5:30 position.

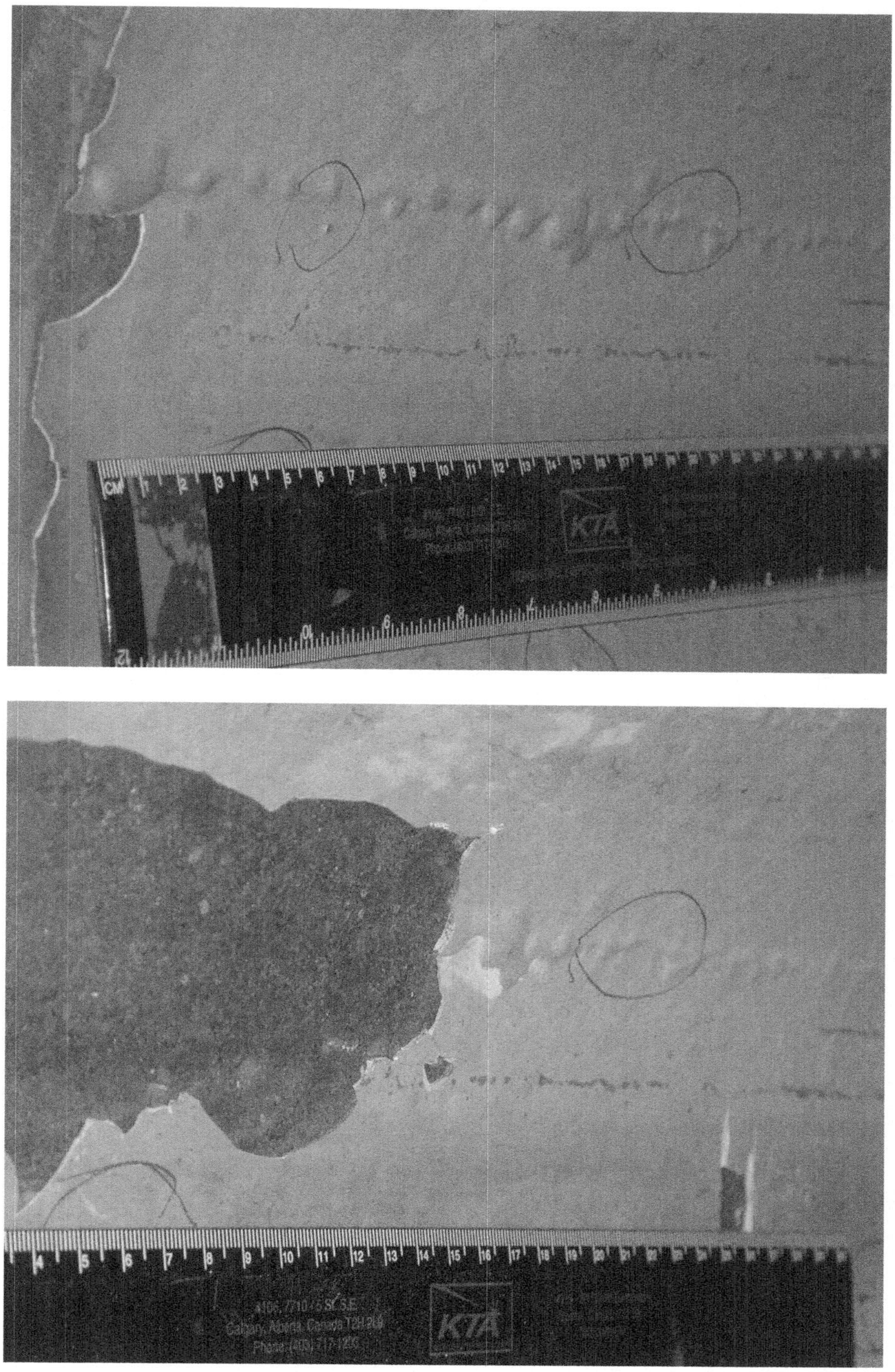

Figure C.32. In sample #106, what looks like a line of blisters is actually dribbles of epoxy lining with some holidays. Lower picture shows the appearance of the steel surface under and beside the dribbles.

SAMPLE #107 – BUCKS COUNTY WATER
(Applied in 2000 [7.5-inch cast iron])

Summary Comments:

- This lining was 4 years old and in excellent condition.
- The lining had no blisters but 7 holidays.
- The film thickness of the lining averaged 49 mils, close to the required 40 mils.
- The finish was glossy with minor bleaching.
- The lining showed considerable ringing.
- The lining had been applied over cement mortar, but retained an adhesion to the mortar, which was similar to the adhesion generally observed for epoxy liners to cast iron; the cement mortar under the lining was in excellent condition.
- Examination of the cement mortar under a holiday showed no visible increase in corrosion of the mortar associated with the holiday.
- Of all of the samples examined, this lining showed the greatest amount of shrinkage away from the pipe wall due to dry out.

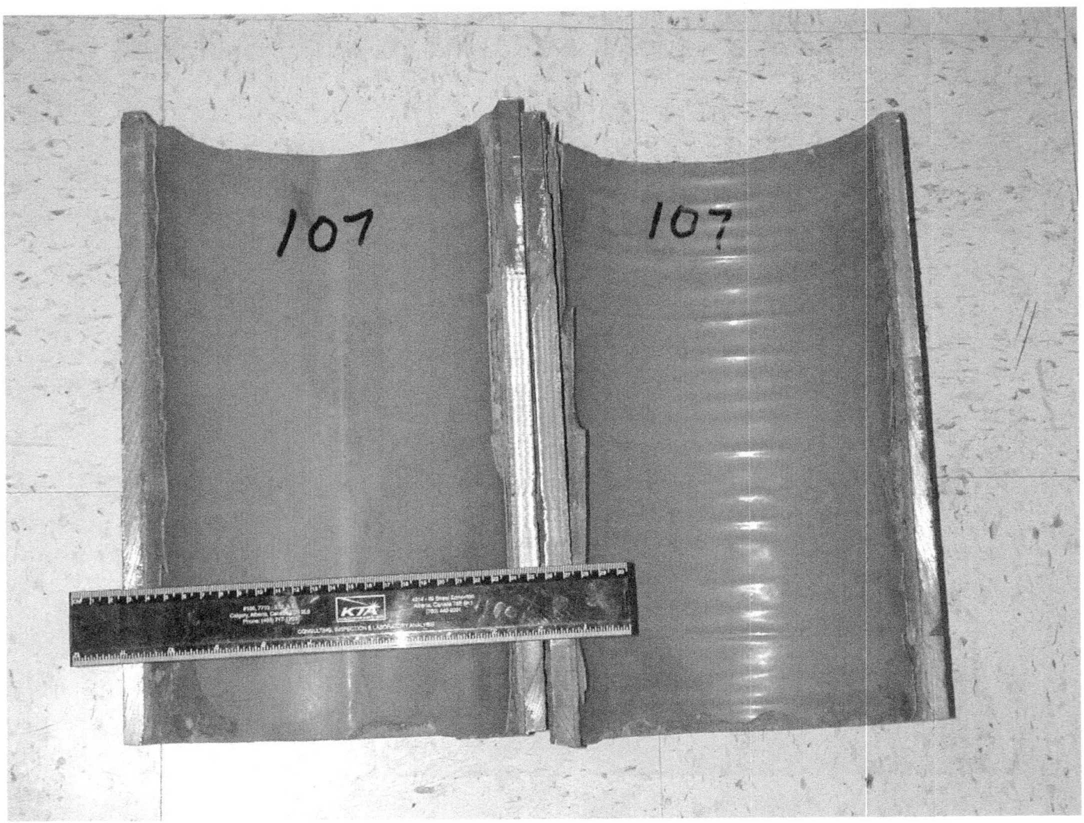

Figure C.33. Sample #107 as received, split in half, showing ringing. Orientation unknown. Note that the pipe was previously lined with cement mortar. Note how the epoxy has pulled away from the cement lining.

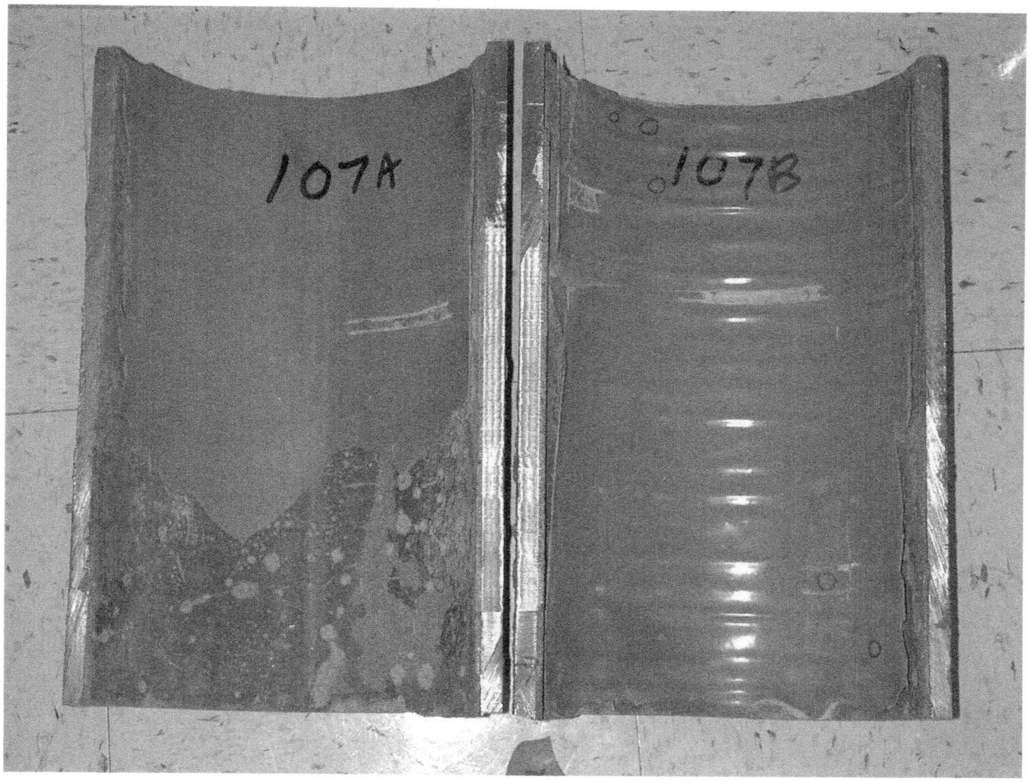

Figure C.34. Appearance of sample #107 after analysis, showing parallel scribe and location of holidays.

APPENDIX D
TEST PROCEDURES OF HISTORICAL PIPE SAMPLES

APPENDIX D
TEST PROCEDURES OF HISTORICAL PIPE SAMPLES

INTRODUCTION

Dry Film Thickness (ASTM D1186)

Dry film thickness was measured on selected panels and compared to the film thickness obtained prior to immersion to identify any changes related to swelling, shrinking, or erosion. The measurement is accurate to ±1 mil. A Positector 6000 electronic magnetic gauge was used, calibrated against National Institute of Standards and Technology (NIST) standards for 0 and 35.6-mil thicknesses. No correction for the roughness of the blast-cleaned surface was applied.

Knife Adhesion (ASTM D6677)

Changes in the adhesion of the lining due to immersion were evaluated by knife adhesion and pull-off adhesion (described below). The knife adhesion method consisted of making cuts through the lining to the base metal, followed by prying the lining with a utility knife to assess adhesion. The adhesion is rated on a scale from 1 to 10, depending on the difficulty of removal and the size of the chips. Note that in this test, the prying force that can be applied may vary with the physical properties of the linings; i.e., a higher force can be applied to a strong lining compared to a physically weak lining where the knife simply cuts through the lining with little prying action.

A parallel scribe cut was used rather than the X-cut described in the ASTM standard. The parallel scribe provided an extended area for adhesion assessment compared to the intersecting arms of the "X." Further, it was easier to assess typical chip size pried from the parallel scribe. The parallel scribe consisted of two cuts, 3/16 inches apart, through the lining to the base metal. The cuts were made with a Dremel tool equipped with an abrasive disc. A 3/4-inch, breakaway utility knife was used to pry at the linings within the scribe.

The parallel scribe adhesion of the linings was evaluated within 1 hour after the panels were removed from the immersion tank. The rating scale in the ASTM D6677 standard (cited below) was used to rate the adhesion:

Rating	Description
10	Lining is extremely difficult to remove; fragments no larger than approximately 0.8 by 0.8 mm (1/32 by 1/32 inch) can be removed with great difficulty.
8	Lining is difficult to remove; chips ranging from 1.6 x 1.6 mm (1/16 by 1/16 inch) can be removed with difficulty.
6	Lining is somewhat difficult to remove; chips ranging from 3.2 x 3.2 mm (1/8 by 1/8 inch) to 6.3 x 6.3 mm (1/4 by 1/4 inch) can be removed with slight difficulty.
4	Lining is somewhat difficult to remove; chips in excess of 3.2 x 3.2 mm (1/4 by 1/4 inch) can be removed by exerting light pressure with a knife blade.
2	Lining is easily removed; once started with the knife blade, the lining can be grasped with one's fingers and easily peeled to a length of at least 6.3 mm (1/4 inch).
0	Lining can be easily peeled from the substrate to a length greater than 6.3 mm (1/4 inch).

Pull-Off Adhesion

One of the few methods by which the adhesion of a lining can be determined quantitatively and objectively is the "pull-off" adhesion method. The test method is described in ASTM D4541 (Pull-Off Strength of Linings Using a Portable Adhesion Tester). Aluminum loading fixtures called "dollies" are bonded to the lined surface by adhesive glue. The lining is lightly sanded and wiped with solvent prior to application of the glue to aid in the adhesion. A hand-held testing apparatus is attached to the loading fixture and a progressively increasing tensile stress is applied to the dolly until separation occurs. The surface is inspected to determine if separation occurred between the metal surface and lining (adhesive failure), within the body of the lining (cohesive failure), between adjacent coats of lining (intra-coat adhesive failure), or between the glue and the lining or fixture (glue failure). The pull-off strength at separation and the nature of the separation are reported. The pull-off fixture used in this program was the Hydraulic Adhesion Test Equipment (HATE) Mark VII, in which force is applied to the dolly hydraulically through a self-centering mechanism. The maximum pull-off strength that the instrument can measure is 2,500 psi.

Pull-off adhesion was determined on the lined panels after the panels were allowed to dry for 1 week after removal from the immersion tank. This ensured that a good bond was obtained between the lining and the epoxy glue. The pull-off adhesion was determined on the blast-cleaned side of the intact panel. In accordance with the ASTM standard, the lining was not cut through to the base metal around the dollies. Three pulls were performed for each lining, and the average reported.

Undercreep Around a Holiday (ASTM D1654 Modified)

Holidays allow direct access of water to the metal surface, an effect that typically leads to the loss of adhesion around the holiday accompanied by corrosion of the metal surface under the lining. The effect is referred to as undercreep, where metal corrosion is often accelerated through crevice corrosion.

Undercreep was evaluated on test panels in which a holiday had been deliberately made in the lining prior to being placed into immersion. The holiday was made by drilling a $^{1}/_{16}$ inch diameter hole through the lining to base metal. Upon removal from the immersion tank, six radial cuts, 60 degrees apart and centered on the holiday, were made through the lining to the base metal. A $^{3}/_{4}$-inch, break-away utility knife was used to pry away the lining next to the holiday so that the metal surface around the holiday could be assessed for corrosion. Specifically, the surface was examined for discoloration and formation of corrosion products. If these were not observed, the undercreep was designated to be zero.

It was not feasible to determine the loss of lining adhesion around the holiday due to the variations in lining thickness, mechanical strength of the different linings, and the subjective nature of the prying action at the "tip" of each sector. As a comparison, the lining was also pried along the radial cuts, at a location well away from the holiday.

Corrosion of the Steel Surface Under the Lining

Water diffuses through all polymer linings to the metal substrate. Depending on the adhesion of the lining and the reactivity of the steel, the water may react to varying degrees with the metal to form corrosion products. It was anticipated that with extended time, visible amounts of bulk corrosion products would form under the coating film. Further, oxygen can also diffuse through the lining, further promoting the corrosion process.

The amount of rusting or formation of visible corrosion products was rated using ASTM D610 (Evaluating Degree of Rusting on Painted Steel Surfaces). ASTM D610 consists of a visual comparator that uses a rating system with a combination of the percentage of the surface corroded and the pattern of the corrosion. The rust grade is rated as follows:

Corroded Area Rating	Description
10	Less than or equal to 0.01% of the surface
9	Greater than 0.01% and up to 0.03%
8	Greater than 0.03% and up to 0.10%
7	Greater than 0.10% and up to 0.3%
6	Greater than 0.3% and up to 1.0%
5	Greater than 1.0% and up to 3.0%
4	Greater than 3.0% and up to 10.0%
3	Greater than 10.0% and up to 16.0%
2	Greater than 16.0% and up to 33.0%
1	Greater than 33.0% and up to 50.0%
0	Greater than 50.0%

A letter rating is added to the numerical rating to take into account whether the corrosion occurs in spots (S), is general over the entire surface (G), or occurs as pinpoint rusting (P).

Blistering (ASTM D714)

Blistering was rated based on the size and density of the blisters in the pictorial standard in ASTM D714. A guide to the pictorial standard, prepared by KTA, is summarized as:

ASTM blister size	mm size	ASTM blister size	% blistered area
10	0 (none)		
8	0.5	F - Few	5
6	1	M - Medium	15
4	2	MD - Medium dense	50
2	5	D - Dense	75

Buchholz Hardness

The objective was to monitor the hardness of the respective linings during immersion and to determine if changes in hardness could be used to assess life expectancy. The Buchholz hardness of the lining film was measured as a function of immersion time, as described in ISO 2815. It was anticipated that hardness would change with immersion time through further cure, impeded cure, embrittlement, or plasticization from hydration. The hardness was measured 1 hour after the panels were removed from the immersion tank and had reached room temperature.

The Buchholz hardness was measured with a commercially made tester. Hardness was determined on the intact, flat panels, on the blast-cleaned side. The method consisted of placing the small indenter (30-mm diameter wheel, 5-mm thick with a 120° bevel along the wheel edge) on the lining surface for 30 seconds under a standard weight of 500 grams (g). The indentation length is measured in millimeters (mm) at 20X magnification within 10 seconds after the indenter and weight are removed from the surface. The harder the lining is, the less the indentation. The mm length of indentation is reported as the result.

Lining Impedance Measurement

The impedance (EIS) of the linings was determined under immersion conditions as a function of the same time and temperature used in the immersion test. The impedance measurements served as a measure of the barrier properties of the linings, and their permeability to water. As linings age, their barrier properties normally decrease due to physical and chemical deterioration, which is accompanied by increasing permeability. It was anticipated that the impedance measurements could be used to evaluate the rate of chemical and physical deterioration of the linings as a function of time, and thus, relate life expectancy to permeability.

The impedance of each lining was measured as a function of immersion time and temperature. Acrylic tubes, 1.5 inches in diameter, were cemented to flat-lined panels. The tubes were filled with tap water to provide the immersion environment. Test panels so prepared were placed in a refrigerator (8 °C), on the laboratory bench (25 °C), or in an oven (35 °C) to obtain the required temperature. The laboratory temperature was controlled at 25 ± 2 °C. Monitoring showed the temperature within the refrigerator and oven varied by less than ± 2 °C. A schematic drawing is shown in Figure D.1.

EIS measurements were made by placing the panel in a Faraday cage, attaching the working electrode to the steel panel, and placing a platinum counter electrode and saturated silver/silver chloride reference electrode (at 25 °C) in the cell. A Bode scan (log impedance versus log frequency) was immediately commenced. Individual panels at 8 °C and 35 °C were quickly removed from the respective freezer or oven for measurement, and then immediately returned. The impedance was determined for the area of the lining enclosed by the plastic tube. This allowed exactly the same area of lining to be measured each time and avoided any disruption of the immersion conditions.

All impedance measurements were normalized to 1 square centimeter (cm^2) of surface area. All measurements were made in duplicate.

Electrochemical instrumentation consisted of the Gamry Corrosion Measurement System PC4/750 potentiostat with EIS 300 software. The root mean square (rms) voltage for the EIS measurements was 500 milliVolts (mV). Bode plots were run from 10^5 Hertz (Hz) to 0.04 Hz. The impedance at 0.01 Hz was read from the graph by interpolation and used in the data analysis.

Wear Test

The objective of this part of the project was to assess the wear resistance of the linings to water flow and scour from fine particulates in the water. The test method was ASTM G6

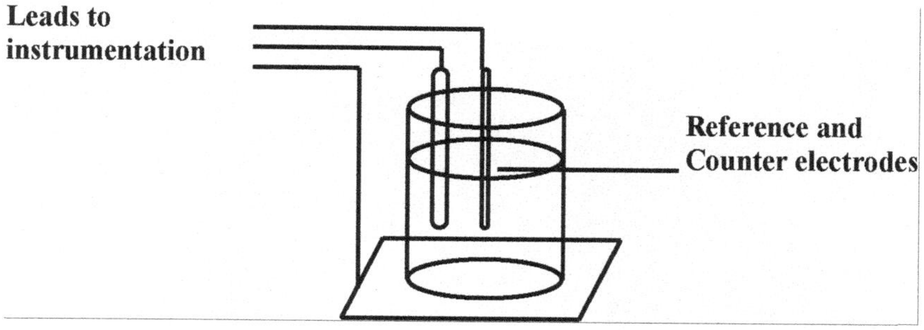

Figure D.1 Schematic drawing of the attached cell for EIS measurements

(Abrasive Wear of Pipeline Linings), which evaluates the wear resistance of linings to slurry abrasion. Various modifications were made to make the test method more representative of potable water erosion. The modifications consisted of:

- Using an abrasive slurry, which was more representative of the potable water system
- Adjusting the test duration so that differences in wear among the epoxies could be discerned
- In addition to film thickness loss, evaluating wear of the linings based on weight loss
- Using loss of electrical resistance solely to monitor the progress of wear during the test, and not as an assessment of amount of wear

The test specimens consisted of lined steel tubes, 0.78 inches in outside diameter and 16 inches long. A photo of the apparatus is shown in Figure D.2. The test procedure consisted of the following. The lined tubes were inserted through the ends of the drum equipped with insulated fittings. The drum, 12 inches long and 14 inches in diameter, was filled with a slurry of tap water (5.68 liters [L]) and abrasive (13.6 kilograms [kg] 24 grit Alodur 24 A/O 3512-8*). The abrasive consisted of aluminum oxide that was the size of common beach sand. The mineral and size were selected as a means to accelerate the test so that wear results could be obtained in a reasonable test period. The drum was rolled on a rolling mill at a rate equivalent to 100 ft/minute at the periphery of the drum. The test was conducted at room temperature (25 °C).

The test was continued until a reasonable distribution of wear was observed among the various linings. The wear was assessed daily as the test was in progress, based on the loss of electrical resistance of the respective specimens. The duration of the test, accordingly, was 11 days. The test was done in duplicate.

The loss of film thickness and weight was determined. The values reported for loss of film thickness refer to specific reference points with maximum wear on each specimen. The wear around the circumference of the tube was not uniform, with most of the wear occurring on the two-thirds of the circumference that faced the center of the drum.

* From Sil Industrial, Edmonton, Alberta, Cana.

Figure D.2 Wear test apparatus

REFERENCES

AWWA (American Water Works Association). 1999. $325 Billion for Pipes. *Mainstream,* 43(2):3.

AWWA (American Water Works Association). 2001. "Rehabilitation of Water Mains." In *Manual of Water Supply Practices, M-28.* Second Edition. Denver, Colo.

AWWA (American Water Works Association). 2002. *AWWA Standard for Spray Applied In Situ Epoxy Lining of Water Pipelines—Draft.* ANSI/AWWA CB1X-xx. Denver, Colo., AWWA.

AWWA (American Water Works Association). 2003. Request for Proposals Life-Cycle Analysis of Epoxy Lining (RFP 2941).

Bhangale, S.M. 2003. Epoxy Resins [Online]. Available: <http://sunilbhangale.tripod.com/epoxy.html>. [cited May 19, 2003]

Canfer, S.J. 1990. *Accelerated Aging Testing of Mercol Epoxy Resin,* Report UC 898,

Conroy, P.J., D.M. Hughes, and I. Wilson. 1993. *Demonstration of an Innovative Water Main Rehabilitation Renewal Technique: In Situ Epoxy Lining*. Denver: AWWA.

DWI (Drinking Water Inspectorate). 2005. *List of Approved Products and Processes.* Wales: DWI.

Gray, L.G.S. and B.R. Appleman. 2003. EIS–Electrochemical Impedance: Spectroscopy: a Tool to Predict Remaining Coating Life. *Journal of Protective Coatings & Linings*. February.

Marcuccio, F. and P. Salvo. 2000. Epoxy Resin Lining: An Eastern Canadian Experience. *Trenchless Technology–Water,* June: 50-52.

NSF (National Sanitation Foundation). 2004. Water Distribution Standards [online]. Available: <http://www.nsf.org/business/water_distribution/standards.asp?program=WaterDistributionSys>. [cited June 15, 2005]

Salvo, P. 1998. In-Situ Watermain Rehab with Epoxy Resin Lining. *Trenchless Technology-Water*. W20-W22.

SSPC (Steel Structures Painting Council). 1990. Performance Testing of Marine Coatings: New Test and Evaluation Procedures. SSPC Publication #90-02 (NSRP0311).

Walker, D., and S. Guan. 1997. Protective Linings for Steel Pipe in Potable Water Service. In *Proc. of the NACE Northern Area International Conference Corrosion Prevention '97.* Toronto, Ontario. NACE.

Warren Associates (Warren Associates (Pipelines) Ltd.). 1995. *Epoxy Lining Consortium: Assessment of Epoxy Resin Linings Used for Water Mains Rehabilitation Long-Term Durability Studies.* 8 Prince Maurice Court, Hambleton Avenue, London Road, Devizes, Wiltshire SN10 2RT, England: Warren Associates (Pipelines) Ltd.

Warren Associates (Warren Associates (Pipelines) Ltd.). 1996. *In Situ Epoxy Resin Linings for Potable Water Applications: Performance Testing.* 8 Prince Maurice Court, Hambleton Avenue, London Road, Devizes, Wiltshire SN10 2RT, England: Warren Associates (Pipelines) Ltd.

Warren Associates (Warren Associates (Pipelines) Ltd.). 1997. *Yorkshire Water Services Ltd. Report on Epoxy Resin Lining Assessment Western Area* (Vol. 1 of 3), First Draft. 8 Prince Maurice Court, Hambleton Avenue, London Road, Devizes, Wiltshire SN10 2RT, England: Warren Associates (Pipelines) Ltd.

Warren Associates (Warren Associates (Pipelines) Ltd.). 2000. *In-Situ Rapid-Setting Polymeric Lining: Operational Guidelines and Code of Practice.* England: Warren Associates (Pipelines) Ltd.

Warren, I.C. 1983. Factors Influencing the Choice of Materials for Epoxy Resin Renovation of Water Mains. *Chemistry and Industry*. 5 September.

Warren, I.C. 2004. Personal communication between I. Warren and L. Gray. January 2004.

Warren, I.C. 1989. In Situ Epoxy Resin Lining — Operational Guidelines and Codes of Practice. Water Research Center, U.K.

Warren, I.C. and S. DeRosa (Warren Associates (Pipelines) Ltd.). 1995. *Durability of In-Situ Applied Epoxy Resin Coatings for Potable Water Mains*. Prague, Czech Republic.

Warren, I.C. and S. DeRosa (Warren Associates (Pipelines) Ltd.). 1996. Epoxy Resin Lining of Potable Water Mains–A Cost Effective Non-Structural Lining Method. In *Proc. of Third National Australian Society for Trenchless Technology Conference*. Perth: Australian Society for Trenchless Technology.

WRc (Water Research Centre). 1989. *In Situ Epoxy Resin Lining—Operational Guidelines and Code of Practice.* England: WRc.

WRc (Water Research Centre). 1997. *In Situ Epoxy Resin Lining—Operational Guidelines and Code of Practice.* Second Edition. England: WRc.

WRc (Water Research Centre). 2001. *In Situ Epoxy Resin Lining—Operational Requirements and Code of Practice.* Third Edition. England: WRc.

WRc-NSF. 2004. Construction Product Testing and Certification: American Materials Testing – NSF 61 [Online]. Available: <http://www.wrcnsf.com/NSF61.htm.> [cited April 12, 2004].

LIST OF ABBREVIATIONS AND SYMBOLS

ac	alternating current
AF	acceleration factor
ANSI	American National Standards Institute
ASTM	American Society for Testing and Materials
AWWA	American Water Works Association
AwwaRF	Awwa Research Foundation
BCWSA	Bucks County Water and Sewer Authority
°C	degrees Celsius
CCTV	closed circuit television
cm^2	square centimeter
DFT	dry film thickness
EIS	electrical impedance spectroscopy
EIS	electrochemical impedance spectroscopy
g	grams
HATE	Hydraulic Adhesion Test Equipment
HRWC	Halifax Regional Water Commission
Hz	Hertz
kg	kilogram
km	kilometer
L	liter
m	meter
mg/L	milligrams per liter
mm	millimeter
MSDS	Material Safety Data Sheet
mV	milliVolts
NACE	National Association of Corrosion Engineers
NIST	National Institute of Standards and Technology
NSF	National Sanitation Foundation
psi	pounds per square inch
PVC	polyvinyl chloride
QA	quality assurance
QC	quality control

rms	root mean square
SSPC	Steel Structures Painting Council
UL	Underwriters' Laboratory, Inc.
WRc	Water Research Centre

IWA Publishing's authorised EU representative for General Product Safety Regulations is Diane D'Arras, 15 rue Duret, 75116 Paris, France, e-mail: safety@iwap.co.uk.

Printed and bound by CPI Group (UK) Ltd, Croydon, CR0 4YY
17/09/2026
02218802-0018